Carbon is Life

Why humans and wildlife depend upon carbon dioxide nutrient,
and how false global warming claims put all our lives at risk

by Ron House,
honorary Australian Magpie

Ron House is a trained physicist and a computer science lecturer, who
decided to investigate the global warming theory after retirement from
full-time teaching. He is a spiritual inquirer with a keen interest in
philosophy, a combination which led him to discover the ethical theory
which he called the Principle of Goodness. He and his wife Gitie have a
special relationship with a family of Australian magpies and their
butcherbird, currawong, and noisy miner friends. Ron has authored
books and papers on computing, ethics, philosophy, humanities, and wild
birds. Ron's websites include: PeaceLegacy.org, PrincipleOfGoodness.net,
and WingedHearts.org.

Carbon Is Life

carbonislifebook.com

Copyright © 2013 by Ron House

ISBN-13: 978-1484919705
ISBN-10: 148491970X

Bunya Grove Press
bunyagrovepress.com

Cover design by the author
containing original photographic art by Karen Hannay,
image elements from istockphoto,
and public domain photograph by Wikimedia user Webaware.
Used by permission.

Table of Contents

Introduction 7

1
Why is Carbon Life? 13
 Carbon Is the Miracle Element in Every Organic Compound 13
 All Life Comes From Carbon Dioxide 17
 When Did Life Arise on Earth? 18
 More CO2 Helps Plants Survive Hardships 23
 Plants Grow Faster and Better with More CO2 23
 CO2 Concentrations Are Near Their Lowest Ever 24
 But What about 'Global Warming'? 27
  The Case for the Prosecution 28
  Preliminary Thoughts for the Defence 33

2
Is Global Warming the Disaster They Say It Is? 39
 The Belief that Warmth Is Bad Contradicts Everything We Know
 from History 42
 Why Should Today's Temperature Be the Perfect One? 44
 What about the Bad Effects of Global Warming? 47
  Rising Sea Levels: 47
  Acidification of the Oceans: 52
  Habitat Loss: 55
 Most of the Warming Would Be in the Coldest Places 57
 Claims about Increased Storms and Hurricanes Are Baseless 58
 Claims about Droughts and Floods Are Baseless 60
 Claims about More Diseases such as Malaria Are Baseless 63
 Polar Bears Are Not Drowning 64
 Claims of "Tipping Points" Don't Pass Reality Checks 66
 Expect Baseless Scare Stories to Keep on Coming 69
 To Summarise... 70

3
Even if Warming Is Bad, Is It Happening? 73
 The Extent of the Warming Has Been Exaggerated 74
 The Hockey Stick Graph 79

Weather Is Not Climate!   83

4
Is Carbon Dioxide a Major Cause of Warming?   87
    The Case Against CO2   89
    Failed Predictions—Warming Theory Refuted   90
        The Atmospheric Hotspot   90
    The Physics of Greenhouse Gasses   99
    The Models Falsified—The Theory Refuted—Again   102
    And the GW Theory Refuted—Yet Again   104
    Problems with the Models Keep on Coming   105
    Hard Data Reverses the Temperature-CO2 Link   108
    Statistics Rips the Climate Models to Shreds   108

5
Lots of Things that Have Been Overlooked   111
    Is the CO2 Concentration Even Increasing?   111
    And in Fact There Are Other Explanations   113
        Other Human Activities   113
    Cosmic Rays and Other Natural Causes   115
    Svensmark's Supernova Revolution   118
        Understanding the Supernova Theory   119
        What Does It All Mean?   121

6
Why Do They Demonise Carbon Dioxide?   125
    Conspiracy?   130
    ClimateGate   130
    The Motivation?   131
    Our Prospects   141

7
What Is the Real Climate Crisis?   143
    A Killer Ice Age   143
    The Death of Everything   146
    Have We Already Saved Earth From Icy Disaster?   148
    And We Face Many Other Deadly Threats   152
    Other Human-caused Influences   155
        Tree Clearing   155
        Nitrogen from Over-fertilisation   156

Subdivision of Natural Wildlife Areas 156
A Myriad Other Dangers 159
Maybe There's Just Too Many People? 159
A Sober Assessment 160

8
The Precautionary Principle 163
The Cautionary Tale of Malaria and DDT 164
Faulty Precautions in Climate Politics 169
Wind Turbines: Wildlife Killing Machines 170
Consequences 172
Which Precautions Make Sense? 173
How Do We Take (Really) Effective Precautions? 176

9
Benefits of Carbon Dioxide 179
Global Warming—Yes, a Benefit! 179
Resilience to Global Cooling 180
Enhanced Plant Growth 181
Plants Cope Better with Environmental Stress 184
The Planet is Turning Green! 185
How Will We Feed 9 Billion People? 186

10
Saving Ourselves, Wildlife, and the Planet 189
Don't Confuse Renewable Energy with Climate 190
Reject All Schemes to Restrict Carbon Emissions 192
Stop Buying into Scare Campaigns 193
Oppose the Use of Ethanol for Fuel 193
Scientific Research 194
Independence for Scientific Research 196
Stand Up for the Web of Life 197

11
Amidst the Chaos:
Seeds of a Higher Civilisation 201
What (if Anything) Lies Beneath? 202
It Has Happened Before (and It Wasn't Pleasant) 204
Visionary Findings—or Finding a Vision 207
Some Suggestions 210

  Suggestion I: 210
  Suggestion II: 211
  Suggestion III: 213
  How do these ideas help us?   213
Synergy of Goodness   215

Appendix
Resources   221
  To Follow the Debate   221
  Web Sites   222
  Books   224
  The Other Side   227
  Finally...   227

Illustration Index

Figure 1: A lone hydrogen atom has an incomplete electron shell   15

Figure 2: Two hydrogen atoms share their electrons   15

Figure 3: A typical carbon-based miracle (sucrose molecule)   16

Figure 4: All life on Earth comes from carbon dioxide   18

Figure 5: Life has progressively drained carbon dioxide from the atmosphere   25

Figure 6: Earth's temperature is usually about 10C hotter than it is today   46

Figure 7: NASA's own graphic shows long-term sea level rise   49

Figure 8: Sea level rise is not accelerating   50

Figure 9: Sea levels at longitude 318, latitude 26   50

Figure 10: Hurricane energy has been flat for 60 years   58

Figure 11: The number of strong to violent F3-F5 tornadoes is also flat   59

Figure 12: Long-term drought records show no worrying trends.   61

Figure 13: U.S. percentages of areas wet and dry since 1900   62

Figure 14: Precipitation in the contiguous United States since 1895 – a small but welcome upward trend   62

Figure 15: Number of temperature stations and the Earth's supposed sudden warming   75

Figure 16: GHCN Raw and Adjusted Temperatures, Darwin Airport   76

Figure 17: The Hockey Stick Graph   79

Figure 18: Fred in bed   91

Figure 19: A blanket holds hot air close to Fred.   92

Figure 20: With no blanket, warm air escapes and Fred shivers   92

Figure 21: Warm air collects in a contained region, so there must be a blanket   93

Figure 22: Warm air escapes upwards, so we can be sure there is no blanket   94

Figure 23: Atmospheric temperature changes as predicted by a climate model ('+' shows warming, '-' shows cooling; period 1958 – 1999). Note the tropical 'hotspot'.   96

Figure 24: Measured data shows a completely different temperature trend, which has no hotspot. 97

Figure 25: Every model gets the flux the wrong way round 104

Figure 26: North American Winter Snow 145

Figure 27: The past four of dozens of repeating ice-age cycles over the last 2.5my. 146

Index of Tables

Table 1: First three rows of the Periodic Table of Elements 14

Table 2: pH values of some common substances 52

Table 3: Atmospheric CO2 Background 1826-1960 shows higher values than the commonly accepted figure 280 ppm. 112

Introduction

The cycle of life is a precious gift
For five billion more years, our good Sun will radiate upon the Earth
The plants will take of the Sun's energy abundance
And the water of the Earth
And the carbon dioxide nutrient in its atmosphere
And create food for animals and for humans, who
Will then replenish the atmosphere with CO_2 so the cycle may continue.
But now war has been declared
Against life itself
The very food of life is called "pollution"
Planetary warmth, which has brought abundance and profusion of life
At every time in the past when it has occurred
Is now demonised, a monster story told to children.
The pale, deathly ice, killer of the Southern continent
Killer of the northern Islands
One-time killer of life over the entire world
Is now worshipped, its every minor retreat bemoaned
Yet, strangely, its advances are kept secret
As the human species, shunning its duty
To become the intelligence and protector of the Earth
Steadily goes insane
Grimly intent to wreck its own prosperity
And to kill human, plant, and animal life
In worship at the altar of a false religion of hatred
Against humanity and against carbon
The essential building block of life on Earth.

Look in the mirror: you are looking, to a first approximation, at nothing but carbon dioxide and water. You, me, that pot-plant, your pets, the grass outside, the food you eat, your friends and family, the forests, the prairie grasses, all the Earth's wildlife, every living thing on Earth—is almost entirely carbon dioxide and water. To be precise, about 93% to 94% of your body weight started out as CO_2 and H_2O, and was converted by photosynthesis in plants into glucose ($C_6H_{12}O_6$) and oxygen. The glucose was then used to construct all the organic molecules in your body, whilst the oxygen, which we later breathe and 'burn' for energy keeps us alive.

These were some of the fascinating things I learned as a science-intoxicated child in the '50s and '60s. I wanted, for as long as I can remember, to know 'what makes things tick'. How does the universe work? Why do we exist? What makes some things 'alive' while other things are not? — You name the question, chances are I asked it. Years later, this 'need to know' meant that the only subject I could possibly study at university was physics. At the core of scientific knowledge, the universal physical laws are the foundation on which all other scientific knowledge is built. I was privileged to study quantum mechanics at post graduate level under one of the last pupils of the great Erwin Schrödinger, an experience for which I have been grateful ever since.

The course of my life, however, determined that it should be in computer science that I would make my daily bread. Physics may be the key to understanding the physical universe, but computing, as well as being great fun, is in more demand. But it tends to absorb one's time, which might explain why it was not until I left my computer science lecturing position that it dawned on me that I had allowed the entire furore about global warming to blow up without my knowing the basic facts about it.

That was in August 2008. By the end of September 2008 I was shocked by what I had learned. I discovered that "global warming", shorthand for the idea that humans were causing dangerous planetary warming by industrial emissions of carbon dioxide, was not only fallacious, it was fraudulent. We were led to believe that we were protecting the environment by "fighting carbon pollution", "reducing our carbon footprint" and so on. But the truth is that our society and our governments have become committed to policies that attack the

Carbon is Life

carbon cycle, the process of plants turning carbon dioxide into food, which is the basis of life on Earth.

That is a very serious charge, that our leaders have "sold us down the river" with policies that seriously damage the planet whilst superficially appearing to be helpful. If true, it means that everything we as a society are now doing to protect our planet is having the reverse effect: we should be *increasing* our carbon footprint (more properly called our carbon donation) if we really want to protect wildlife, save the wilderness, and safeguard our own children's futures. It is hard to imagine a more important question to which we should find the right answer—that is, the truth, not what makes us feel good.

The elements of the carbon cycle as I described it in the first paragraph are known, indisputable scientific facts. On the other hand, as I learned to my horror, the alarm about "global warming" is based entirely on computer models that are *already known to be wrong,* and *nothing else.* There is certainly no actual evidence for it, not one tiny speck. Carbon dioxide is an indispensable part of the cycle of life.

Carbon dioxide is the food of life—it is 100% good. More CO_2 makes plants grow better, faster, with less water. It is used in professional greenhouses at very high concentrations for exactly that reason. If the atmosphere contained more of it (even a *lot* more of it), it would insure the world's poor against starvation. But instead, politicians and others are running a fraudulent scare campaign against it and doing evil things like using food for car fuel, which has already doubled food prices and thrown millions of the most desperate into starvation.

No issue could be more important. Stopping a fraud that threatens the very basis of life itself is the most urgent challenge facing us all. Here are some basic facts that everyone really, really needs to know:

- carbon dioxide (CO_2) is close to its *lowest* concentration *ever* in the entire 4.3 *billion*-year history of planet Earth;
- *more* CO_2 increases the growth rate of crops, wild plants, trees, the food of all animal species; but the Earth's plants are close to carbon starvation; calling CO_2 "pollution" is a deadly

mistake;

- given our planet's current (much *colder* than normal) climate, global warming would be good for life, humanity, and the environment;
- but unfortunately carbon dioxide's only failing is—it's not causing much warming!
- and the only realistic global climate temperature danger we face is a major planetary cooling;
- the vastly expensive cap-and-trade systems and carbon taxation being introduced will wreck the economies of the West for no benefit;
- and they will reduce the food for wild animals and for the poor, and make it slightly more likely the planet will enter a deadly ice age, killing billions of humans and animals;
- the most urgent thing anyone who cares for the environment can do is help stop the deadly anti-carbon juggernaut.

Unfortunately we have largely been kept in ignorance of some basic, but critical, information. Some facts about climate which we all need to know (and I'll show evidence for all of this in the following pages) are:

- Climate changes all the time—and the only "climate change deniers" are the global warming alarmists who tried to pretend that the Earth had static temperatures for one or two thousand years up until the industrial revolution;[1] the truth is that Earth's temperatures has varied over four or more degrees in the past few thousand years alone;

---

[1] Yes, they really claim the climate has been constant! America's NOAA National Climatic Data Center, who have the expertise to know better and who are one of the premier 'scientific' bodies investigating climate, officially wrote this untruth about one of their own scientific reports: "The report emphasizes that human society has developed for thousands of years under one climatic state, and now a new set of climatic conditions are taking shape." This wacky statement ignores the fact that humans live under different climate states, not only from time to time as Earth's temperature rises and falls, but at the very same time—living everywhere from Siberia to the Sahara desert to the Amazon rainforest. http://www.noaanews.noaa.gov/stories2010/20100728_stateoftheclimate.html

- The long-term geological history of the Earth over the past four billion years shows no relation between carbon dioxide concentration and temperature;
- Mass extinctions of life can and have occurred when temperature plummets;
- Some very warm periods of the planet's history have been amongst the most prolific and abundant;
- The Earth's climate has been (and possibly has *mostly* been) ten or more degrees Celsius hotter than now, and life has done well;
- Earth has had CO_2 concentrations 2,000% higher than today, and life has flourished;
- CO_2 is a natural fertiliser, and the projected increases over the coming decades can make crops give anything from 30% to 70% greater yield—trying to prevent this amounts to trying to starve people;
- "Clean coal" is an insane attempt to use (roughly) an extra 40% of our precious non-renewable resources in order to power the pumping of carbon dioxide nutrient deep underground, thus denying it to plants and to life, whilst causing us to run out of non-renewable resources even sooner;
- "Cap and trade" and other carbon emissions reduction schemes are designed to make you poor while they deny Earth the benefits of carbon dioxide nutrient and a more life-supportive planet;
- Furthermore, the money trail shows that 'Big Coal' actually supports the global warming hoax—after all, if you need an extra 40% coal to deny life to animals and humans, then that 40% must be provided by—you guessed it— 'Big Coal'!

The shocking story of the hows and whys of humanity's sudden lurch into insanity—an insanity that threatens almost every species of life on Earth—is too large to tell in full here. My mission in this book is only this: to show you that it is indeed happening, that we need to change course if we are to stop horrific harm to the environment, to the creatures that share our planet with us, and to the most disadvantaged of our own species; and to suggest some actions we can take together to help others see reason and to put our planet on a genuinely creative and life-affirming path.

1
Why is Carbon Life?

Carbon Is the Miracle Element in Every Organic Compound

Carbon has long been recognised as a very special element indeed. Chemistry is divided into two main specialities: organic and inorganic. Organic chemistry is the study of compounds containing carbon, whilst inorganic chemistry is the study of *everything else*. And there is a good reason for calling compounds of carbon "organic": the molecules necessary to make a living being, as far as we know, must be based on the element carbon.

For readers who are not 'into' chemistry, it is still worth taking a short look at just why this miracle element is, indeed, a miracle. And it's not hard to follow the important points. The following is a simplified description of the physics and chemistry of carbon, but it covers the key facts relevant to our topic.

An *element* is just a type of atom. The element hydrogen is those atoms that contain one positively charged particle, or "proton"; helium atoms contain two protons, lithium three, and so on, up to uranium, the largest naturally-occurring element, which contains ninety-two protons. The number of protons is called the *atomic number*. Now normally, atoms have no electric charge, and since the protons have a positive charge, the atom must contain the same number of negative charges. For this reason every normal atom also has the same number of negative "electrons" as it has positive protons. For example, atoms of carbon all contain six protons, so they also contain six electrons.

It turns out that the number of electrons an atom normally has is very important in how that element behaves chemically. If you look up "Periodic Table" on the internet, you will see that the elements are arranged in rows. Carbon is in row two, containing the elements with atomic numbers from three through ten. Carbon, number six, is exactly in the middle between element two (helium) at the end of the previous row, and element ten (neon) at the end of row two, as shown in Table 1.

hydrogen 1 proton							*helium* **2 protons**
lithium 3	beryllium 4	boron 5	**carbon** **6**	nitrogen 7	oxygen 8	flourine 9	*neon* **10**
sodium 11	magnesium 12	aluminium 13	silicon 14	phosphorus 15	sulphur 16	chlorine 17	*argon* **18**

Table 1: First three rows of the Periodic Table of Elements

The reason for the rows in the periodic table is that these represent the shells[2] around the atom's nucleus into which the electrons are packed. The elements at the right-hand end of each row are called the "noble gases". These elements almost never chemically combine with other elements because they have their electrons neatly packed so that they exactly fill the spots available in the shells. This makes them very stable and unreactive chemically. At the risk of a spot of anthropomorphism, we can think of the noble gases as being "happy" just the way they are.

The first shell has only two spots, so helium, with two electrons, exactly fills the two spots and the helium atom is "happy". The next shell has eight spots, so neon, with ten electrons, exactly fills all the spots in the first two shells, and neon is also "happy". Likewise argon's eighteen electrons exactly fill three shells and so argon is "happy".

The other elements, we might say, are a bit disgruntled. They would really like to fill their shells with electrons, and there is a way they can do that. By linking to other atoms (which is called *forming compounds*) they can share electrons with another atom so that each

---

2   The shells are in turn comprised of more complicated arrangements for the electrons called orbitals, but that doesn't change anything that concerns us here.

atom seems to have a complete shell of electrons. So, from the table, we see that chlorine, for example, is one electron short; it can therefore make one chemical bond and it will then be able to simulate the completed electron shell of the next element, argon. Likewise, sodium, which is in excess by one can form exactly one bond and simulate the element neon. So a sodium atom, by 'donating' its extra electron to a chlorine atom, can make a single bond with it to form a molecule—ordinary common salt, sodium chloride. One of the simplest molecules, hydrogen, demonstrates molecular bonding most clearly. The first electron shell has two 'slots', but a hydrogen atom only has one electron. So an unattached hydrogen atom has an unfilled electron shell, as shown in Figure 1.

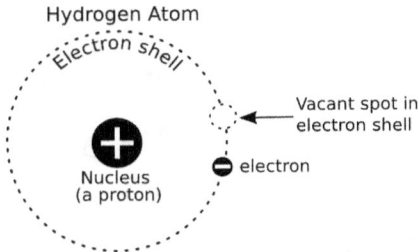

Figure 1: A lone hydrogen atom has an incomplete electron shell

By sharing their electrons, two hydrogen atoms can each 'pretend' to have a complete shell with two electrons, as shown in Figure 2.

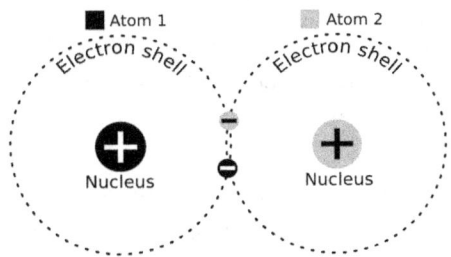

Figure 2: Two hydrogen atoms share their electrons

The second electron shell has eight slots for electrons, which makes things more interesting. Those elements that are two electrons short or two electrons over the magic number that fills the electron shells can make two bonds. This is why oxygen (two short of filling the second shell) can have two bonds with other atoms. Likewise, nitrogen (three short) can have three bonds.[3]

As you can see, the very special element carbon has four more electrons than the magic number two (of helium) and four less than the magic number ten (of neon). This means that carbon is very versatile and can make four bonds, and it will bond to almost anything —including other carbon atoms. That is why carbon can make long chains, creating very complex molecules—alcohols, proteins, carbohydrates, sugars, amino acids—all the ingredients of life (Figure 3).

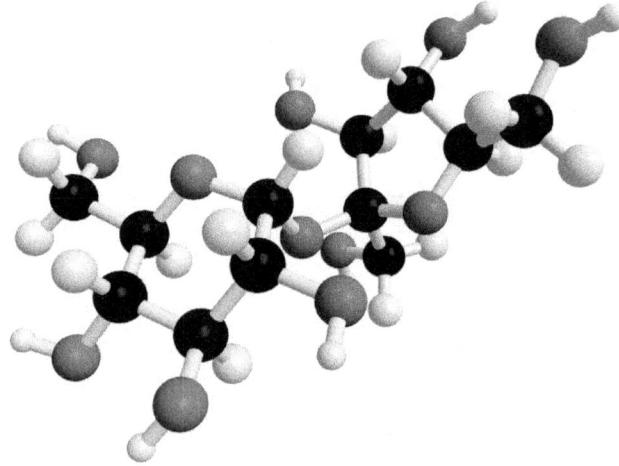

Figure 3: A typical carbon-based miracle (sucrose molecule)

If you look at the three rows of the periodic table shown in Table 1 earlier, you might object that silicon (on the row below and directly under carbon) also has the same properties of having four more electrons than needed to fill the shell below, as well as four too few to fill the next shell. Indeed it does, but there are a number of subtle differences: for example the carbon atom is just the right size to allow its long chains to fold up in many novel ways; silicon is a bit too big. It turns out that how molecules fold is critical to the chemistry of life. Some scientists speculate that somewhere else in the universe,

---

3   This discussion skips over the important distinction between ionic and covalent bonds, but it is sufficient for our current purposes.

life might be based on silicon—but here on Earth, it's all carbon.

All Life Comes From Carbon Dioxide

So how does the carbon get into us? How do we acquire all the carbon needed to make all the molecules of life that keep us alive? The chlorophyll in green plants uses energy from the sun to transform water and carbon dioxide into glucose and oxygen. The oxygen is released into the air; animals, including ourselves, eat the plants, obtaining the essential organic molecules created in the plants, and then we breathe in that same oxygen that was released during photosynthesis, and we combine it with the organic food (in effect undoing the reactions of photosynthesis) to release the stored solar energy and thus power our bodies—as well as re-forming carbon dioxide, which we breathe out to be once again available for forming organic molecules in another plant that will be eaten by another animal—and so the cycle of life goes on.

Carbon, oxygen, and hydrogen—obtained from the two miracle molecules of life, carbon dioxide and water—food and drink for plants—with energy supplied by our good Sun—make up the vast bulk of every living being. These three elements alone make up around 94% of the body weight of a human being. So when the alarmists tell you to fight "carbon pollution", they are telling you to wage war against the very stuff you are made of. Figure 4 shows the major net movements of carbon through the carbon cycle—or should it be *the cycle of life?*

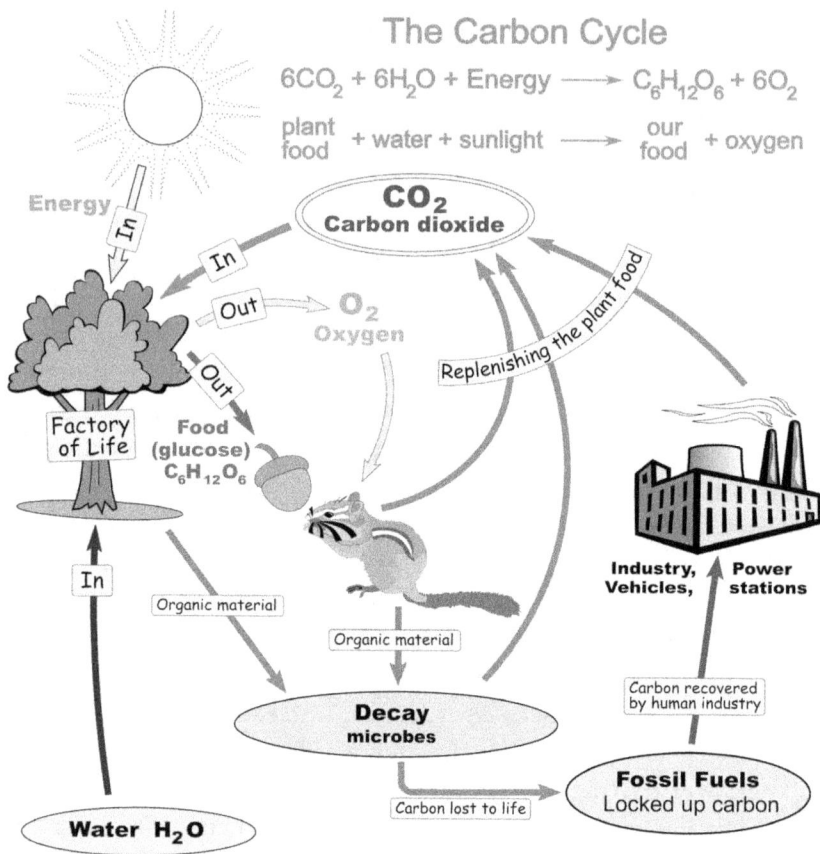

The Carbon Cycle

$$6CO_2 + 6H_2O + Energy \longrightarrow C_6H_{12}O_6 + 6O_2$$

plant food + water + sunlight $\longrightarrow$ our food + oxygen

CO_2 Carbon dioxide

Energy

In

In

Out O_2 Oxygen

Out

Replenishing the plant food

Factory of Life

Food (glucose) $C_6H_{12}O_6$

In

Organic material

Organic material

Industry, Vehicles,

Power stations

Carbon recovered by human industry

Decay microbes

Fossil Fuels Locked up carbon

Carbon lost to life

Water H_2O

Figure 4: All life on Earth comes from carbon dioxide

When Did Life Arise on Earth?

Life arose remarkably early in the history of the Earth. Geologists divide the history of the Earth into various time periods, the longest of which are the *Eons*. Earth was formed about 4.5 billion years ago, and the first eon, the *Hadean eon*, was the time when the Earth was still cooling down from its formation. During this time some terrible events happened, and the Earth, for much of this time, would have been like hell, or hades. There was the suspected collision with a Mars-like body that would have melted the planet, and which resulted in the formation of the Moon. There was also the mysterious *late heavy bombardment*, a 300 million-year period when meteorites rained down upon the planets of the inner solar system; yet even at this time primitive life almost certainly already existed[4]. The Hadean eon ended about 3,800 million years ago (or 3,800 mya for short) when the late heavy bombardment ended, and yet even as early as this, there is indirect evidence of photosynthesis. So life, remarkable life with photosynthesis and all, appeared virtually as soon as it was possible for liquid water to exist on the surface. Thus did the wonderful legacy of the miracle element, carbon, begin.

The first life forms were primitive bacteria, of which the very first were probably the *extremophiles*, which could tolerate high temperatures or acidic water or other conditions that would be fatal to most life today. The extremophiles are still with us, but they have retreated to hot vents in the seas, super-salty waters like the Dead Sea, and other places that 'normal' life cannot tolerate.

The time up until multi-cellular plants and animals arose about 540 million years ago (mya) is called the *Precambrian*. For much of that time a careless extraterrestrial observer might have thought the Earth was dead, but life was there and life was active. Cyanobacteria ("blue-green algae") steadily generated oxygen. At first this oxygen was absorbed (for example, by turning surface iron into rust) but eventually so much oxygen was created that it started to collect in the atmosphere, laying the foundation for more advanced life forms.

A significant event of the Precambrian that concerns us is that, to the best of our knowledge, the Earth froze solid, all the way or nearly all the way to the equator, at least twice. If such an event happened today, it would be the end of multicellular life as we know it—including ourselves. Only bacteria could possibly survive a global freeze, to once again inherit the planet they once ruled.

The Precambrian ended with the appearance of the very first multicellular life forms: the *Ediacaran fauna*, strange forms that we would scarcely recognise as living beings. This was followed by the

---

[4] http://www.planetary.org/news/2009/0529_Ancient_Asteroid_Storm_May_have _Aided.html.

most remarkable period in our planet's history, the *Cambrian explosion*, which initiated the *Phanerozoic eon*, which has occupied the last 540 million years, the time of 'our' life—recognisable multicellular plants and animals. The Phanerozoic is divided into *eras*, which are divided into *periods* chronicling the development of life:

- the *Paleozoic era* (before the dinosaurs, 642 mya – 251 mya), contained these periods:
    - the *Cambrian* period: the time when ancestors of all the modern types of plants and animals came into existence; this is the astonishing time of the Cambrian explosion—the sudden, near-miraculous, flourishing of life-as-we-would-recognise-it in all its wonderful variety;
    - the *Ordovician*: the first green plants and fungi emerge onto land; but this period also contained a glacially cold time that likely caused the second-greatest mass extinction of life in Earth's history, in which 75% of marine species died;[5]
    - the *Silurian*: the first land animals (such as millipedes) appear, along with the first jawed fishes;
    - the *Devonian*: the first mosses and ferns appear, the first seed-bearing plants, the first insects; early sharks and the first amphibians;
    - the *Carboniferous*: large primitive trees; land vertebrates; huge coal-forming swamps; winged insects appear and the first reptiles;
    - the *Permian*: marine life flourishes, beetles and flies appear; the supercontinent Pangaea is formed by continental drift from all the land masses on Earth; trilobites flourish; but this period ends in a holocaust with the greatest mass extinction in the history of the Earth, when 95% of all species went extinct;
- the Paleozoic was followed by the *Mesozoic era* (the time of the dinosaurs, 251 mya – 65.5 mya), also when mammals emerged,

---

5   http://www.sciencedaily.com/releases/2011/01/110127141703.htm?
     goback=.gde_3062307_news_345093490

Carbon is Life

to be overshadowed by their more glamorous cousins until the apocalyptic events that ended the world for those 'terrible lizards' who still fascinate humans today; it is also divided into periods:

- the *Triassic* period: when life struggled to recover from the devastation of the Permian extinction event; the emergence of the great ocean-going ichthyosaurs, the flying pterosaurs, the first mammals and crocodiles;
- the *Jurassic* period: the maturing of the dinosaurs, the time of the massive sauropods, the giant long-necked plant-eating creatures such as Diplodocus and Brachiosaurus, whose abundant leafy diet was provided courtesy of the high CO_2 concentration in the atmosphere; the first birds appear; the supercontinent Pangaea breaks apart;
- the *Cretacious* period: the final great flourishing of the dinosaurs, with even more massive sauropods, the great Tyrannosaurus Rex; my favourite, Triceratops; modern fish; flowering plants; modern sharks; birds replace the pterosaurs; the modern mammalian groups monotremes, marsupials, and placentals appear; the world is warm, very warm, and life flourishes, until the mass extinction that doomed the dinosaurs and gave mammals, in the end including ourselves, their chance to dominate the world;

- the *Cenozoic era* (after the dinosaurs, 65.5 mya – present); whilst most of our planet's history has been warm, the Cenozoic era has slowly become increasingly cold: about 38 mya[6] Antarctic ice began to form[7]; starting about 2.58 mya, Earth cooled enough that it entered an ice age, and it has oscillated into and out of ice ages ever since (each one colder than the

---

6   "mya" — million years ago.

7   Kennett, J. P. (1977), *Cenozoic Evolution of Antarctic Glaciation, the Circum-Antarctic Ocean, and Their Impact on Global Paleoceanography*, J. Geophys. Res., 82(27), 3843–3860, doi:10.1029/JC082i027p03843.
http://www.agu.org/journals/ABS/1977/ JC082i027p03843.shtml

one before);

- most of the Cenozoic is called the *Pleistocene* period, and is the time of emergence of the woolly mammoths, sabre-toothed cats, and all the other mammal megafauna, the earliest humans; and a near planetary cataclysm due to *lack of* carbon dioxide that almost ended life as we know it (which we shall discuss later); it was once thought that this was also the time of emergence of arguably the most advanced life-form on the planet: grass, but we now have evidence that grass emerged much earlier and was actually eaten by some dinosaurs;

- finally, the time since the last ice age ended some 11,400 years ago is called the *Holocene*; in reality, the Holocene is just one more interglacial, a short time of a few thousand years in between two deadly ice ages (the next one is overdue); the previous interglacial, just before the last ice age, was called the *Eemian* interglacial.

In writing this I noticed that the eras of the Phanerozoic are named by their relationship to the dinosaurs. A tear came to my eye for the loss of these amazing animals, until I remembered that our wonderful, beautiful friends, the birds, are indeed actually dinosaurs, who are still with us today.

In reviewing this very short history of life, surely we cannot but be humbled by the richness of the legacy that our wonderful world has bequeathed us. The magnificence of the long-gone creatures—most famously the dinosaurs, but also the huge mammals of the early Cenozoic, such as Baluchitherium, which weighed more than two elephants—as well as the massive trees and abundant plants that fed them, were only possible because our planet's atmosphere then had much more carbon dioxide than it has today. Although it is not obvious, the high CO_2 was essential for the success of all these plants and animals, for reasons we shall examine in the following sections.

More CO_2 Helps Plants Survive Hardships

It's obvious, but it has also been tried, tested, and proved anyway by innumerable scientific experiments, that if you have plenty of food you'll survive other things better than if you are starving in the bargain.

Plants surrounded with more carbon dioxide require less water, so they can grow better in drought conditions. Quite apart from human usage in growing food, this provides more options for wildlife, who are hard pressed from encroachments by people upon their natural habitats.

Plants Grow Faster and Better with More CO_2

Owners of greenhouses routinely add carbon dioxide to the air in the greenhouse to grow more bountiful crops faster. Doubling the CO_2 in the atmosphere increases the growth of plants by from 15% to 40% or even more, depending on the crop.

If atmospheric CO_2 were doubled, we could immediately feed a billion more people without any extra effort, just from the farms that exist today.

Conversely, if CO_2 were to be reduced to the 350 ppm (parts per million) figure that many activists demand, we would immediately lose about ten percent of the world's food supply, enough food to put perhaps 400 million people into starvation. The "350" activists are actually demanding that we starve people!

Even worse, if the CO_2 concentration fell to 285ppm, the figure it is alleged to have been before the industrial revolution, about a third of the world's food supply would vanish, and maybe a billion people or more would starve.

These are not made-up figures; they come from experimentally demonstrated facts about carbon dioxide's effects on the growth of plants.

Again, we should not forget wildlife. Doubling CO_2 would grow more grasses and other food sources for wildlife, providing more ways for wildlife to adjust to and survive human encroachment. Many of us, myself included, are seriously concerned about the pressure on wildlife from the continually expanding human occupancy of the Earth. Population policy, including the question of the optimal human population, is surely a critical issue whichever way our thoughts might lean on the subject. As an intelligent species we need to seriously address the matter because our policy on human numbers influences so many other vital questions. But we should not confuse the two issues of human numbers and protecting wildlife. However many or however few of us there may be, wildlife will always be better off—indeed *much, much* better off—as a result of our emissions of carbon dioxide.

This is the direct opposite of everything we are told by our politicians and the media, but there is no doubt whatever about it. The official message is dangerously wrong; "reducing our carbon footprint" is a huge threat to wildlife because it denies them food, as we shall see in more detail later.

CO$_2$ Concentrations Are Near Their Lowest *Ever*

You wouldn't know it from all the alarm in the mainstream media, but the Earth is now close to the minimum CO$_2$ concentration in its entire 4.5 billion year history. All the plants that now exist evolved on an Earth with much higher concentrations. With only a little less CO$_2$ than we had during the last ice age, plants would stop photosynthesising entirely, and the Earth's plants would die. If humans are now adding to the carbon dioxide in the atmosphere, we have helped the planet avert the worst possible disaster—the end of everything!

Previous times of high carbon dioxide (which is almost *any* time in the history of the world!) have provided the most prolific, life-giving, flourishing environments. Think of the Jurassic, the period when the great herbivore dinosaurs roamed the world. The vast herds flourished only because the bounty of two and a half times more carbon dioxide than today's meagre figure was available to replenish the forest canopies of the giant redwoods and other ancient plant lineages. Or think of the Cambrian explosion, when, in a period of maybe only five million years, all the plant and animal phyla that exist today came into being: the most productive period for life since the Earth was formed 4.5 billion years ago. And the CO_2 concentration was some *twenty times* what it is today, as shown in Figure 5, which shows graphically how life has steadily drained the atmosphere of its CO_2.[8]

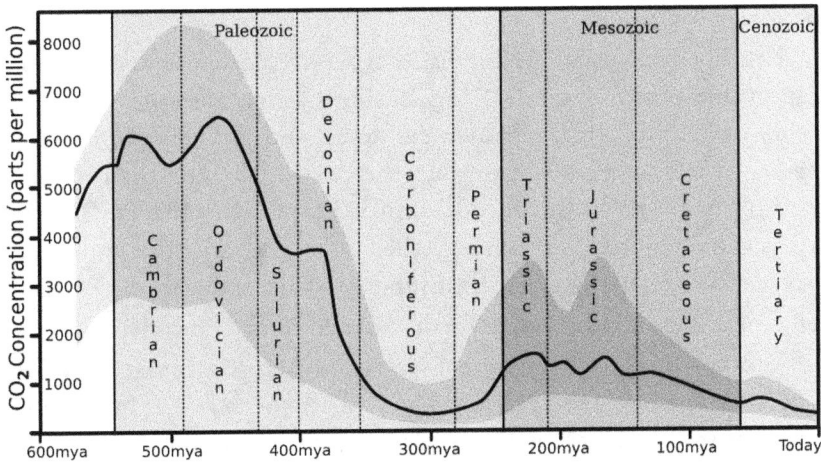

Figure 5: Life has progressively drained carbon dioxide from the atmosphere

The story told in this figure certainly gives food for thought. Is life 'using up' the world's CO_2? If so it would not be the first time that life has changed the atmosphere to its own detriment. Long before the Cambrian explosion, life on Earth was dominated by anaerobic bacteria that produced oxygen as a by-product. Eventually the oxygen built up to such an extent that the atmosphere became poisonous to many of those same life forms (namely the anaerobic bacteria). Only newer life forms that could tolerate the oxygen or use it to 'burn' for energy allowed life to spread across the planet.[9]

---

8   CO_2 values obtained from: Robert A. Berner, The Rise of Plants and Their Effect on Weathering and Atmospheric CO_2. *Science* 25 April 1997: 276 (5312), 544-546. [DOI:10.1126/science.276.5312.544]

9   D.E. Canfield. *The Early History of Atmospheric Oxygen: Homage to Robert M.*

Meanwhile the former inhabitants of Earth, the oxygen-hating bacteria, have been banished to nooks and crannies where the oxygen doesn't penetrate. Does the above diagram warn us of a second holocaust against life—this time, 'our' life, the oxygen-loving life built upon CO_2 plant food? If so, perhaps humans have arrived only just in time to turn around the decline by digging up that buried carbon, locked in coals and oils, and returning it to the atmosphere? If so we have avoided a mass extinction worse than any ever experienced in the planet's history.

This does not mean, of course, that we are doing these things in a responsible way. Typically, big business often tramples over the rights of people, wrecks wilderness areas, and pollutes the environment in the process of digging up the coal or drilling the oil, since their primary goal is profit, not what is best for our planet. But that is a separate question from whether it is a good thing to return locked up CO_2 back to the atmosphere. I have some thoughts about the moral issues raised here, which I discuss in the final chapter.

Another question that might be raised here concerns evolution. Yes, we might say, temperatures have usually been hotter in previous ages, but since then animals and plants have evolved to suit the now-colder conditions. Will today's life still be able to cope if conditions warm up again? As we discuss elsewhere, the predicted warming would happen (if it happens at all) mostly in the coldest places, meaning that, except for the coldest environments, all the environments that exist today will still exist somewhere if the world warms. And the warming of the poles would release a huge area of land for life that is currently iced over and almost, but not entirely, devoid of life. On the whole, then, even for today's life forms, warmer would be better. Yes, a few species (such as polar bears) evolved to make use of very cold environments, but (as we shall see) even for them, their numbers are increasing where temperatures are going up and dropping where temperatures are going down. These species have

Garrels. Annu. Rev. Earth Planet. Sci. 2005. 33:1–36 . doi: 10.1146/annurev-.earth.33. 092203.12271 Online at http://courses.washington.edu/bangblue/Canfield-History_of_Atmospheric_Oxygen-AnnRevEPS05.pdf

evolved, so the evidence shows us, to be able to live in cold places, not to be able to live *only* in cold places.

Lastly, as we consider these huge CO_2 values in previous times, we may indeed wonder why the planet never hit any of the "tipping points" that the alarmists keep warning us about? Was the planet a thermal hell house when the carbon dioxide was at those extreme concentrations? No. It has been cold when CO_2 was high, and warm when it was low. But over the long term, temperatures have typically been some ten degrees Celsius higher than today. We shall look at tipping points more closely later.

But What about 'Global Warming'?

As we shall see shortly, the globe has warmed and the globe has cooled, in cycles of various lengths from a single day (nights are usually cooler than days) to hundreds of years (it is now warmer than it was in the early 1800s) to thousands of years (it is now cooler than it was a thousand years ago) to millions of years (the current era, for the past 2.5 million years, is very much cooler than the millions of years immediately preceding). The climate has changed unceasingly, and no realistic scientist disagrees. But when we hear "global warming" or "climate change" on television and in the newspapers, it is being used as code for a very special idea: that *humans*, by emission of *carbon dioxide*, have caused or will cause a *dangerous* amount of *warming*, with all sorts of projected (bad) consequences. The name given to this alarmist theory is *Anthropogenic* (i.e. human-caused) *Global Warming*, abbreviated to *AGW*, or sometimes even *CAGW*, with a "C" for "catastrophic". Calling it "alarmist" isn't an insult: if it is actually happening, we really should be doing something about it[10]. But is it? This, of course, is the killer question, and we have to devote much of this book to looking at it thoroughly.

---

10 Though not, as it happens, what we are told to do: trying to stop it is a terrible waste of resources that could be used instead to adapt to it, with much more effective results. We won't go into this in depth in this book because the theory is not, in fact, true at all.

The Case for the Prosecution

To me, and to any real scientist, a "case for the prosecution" would consist of evidence of some sort—and by evidence, I mean facts, observations, that build up into a coherent case. But for many, the mere fact that 'everybody' thinks so is taken as evidence. This is one of the classic logical fallacies known for some 2,300 years. As Lord Christopher Monckton explains:

> "...many of the climate-extremists' commonest arguments are instances of logical fallacies codified by Aristotle in his *Sophistical Refutations* 2300 years ago. Not the least of these is the *argumentum ad populum*, the consensus or head-count fallacy.
>
> "The fallacy of reliance upon consensus, particularly when combined with the *argumentum ad verecundiam*, the fallacy of appealing to the authority or reputation of presumed experts, is more likely than any other to mislead those who have not been Classically trained in mathematical or in formal logic[11]."

I raise this before discussing the real evidence because so many of my friends have been incredulous that I could disagree with so many and so powerful people and organisations who are all completely swept up in the alarmism. These range from:

- just about every Western government and mainstream Western political party,
- virtually the entire 'official' infrastructure in these countries (government departments, universities, etc.), and
- a very special international organisation, the IPCC, or in full, the Intergovernmental Panel on Climate Change. It was set up in 1988 by the World Meteorological Organization (WMO) and the United Nations Environment Program (UNEP), to, in their own words, "prepare a comprehensive review and recommendations with respect to the state of knowledge of the science of climate change; social and economic impact of climate change, and possible response strategies and elements for inclu-

---

[11] Christopher Monckton of Brenchley. *Why there cannot be a global warming consensus.* http://wattsupwiththat.com/2012/04/23/why-there-cannot-be-a-global-warming-consensus

sion in a possible future international convention on climate[12]."

The IPCC has prepared four Assessment Reports, becoming increasingly strident in each successive report about the horrors awaiting the world from "climate change". Who could *possibly* disagree with that?

We might expect the answer to be: anyone who, in Lord Monckton's words, understands basic logic. Unfortunately that's not true. Social psychologist Solomon Asch's famous "elevator experiment" proved that the joke's on us: we are ready to do and believe illogical things if enough other people do and believe them all around us.[13] Regardless of the threadbare logic and lack of evidence, we find ourselves overwhelmed by the emotional persuasion of the whole edifice of our society pressing down upon us. How could they all get it so wrong? Maybe because they, like us, only believed it because others already believed it? We take it for granted that if so many already believe it, quite a few of them must have checked it out properly somewhere, sometime. We don't imagine that it could be a small clique of scientists manipulating the scientific peer review process[14], and getting up to antics like deliberately stopping the air conditioning when they are due to explain 'global warming' to a bunch of scientifically ignorant politicians[15]. I'll offer some more thoughts about that in Chapter 6, but let's set that aside for now and just consider the allegations.

What are the major alarmist claims that we shall examine in later chapters? We hear them endlessly on television and in the papers (the mainstream media); here are some highlights:

- of course there is the big one: *global warming*, aka 'climate change', aka 'global climate disruption', aka 'climate challenges' —the name keeps changing for an insidious reason

---

12 http://www.ipcc.ch/organization/organization_history.shtml#.UNqXNpZ1VhE

13 Do a search for 'elevator experiment' to see a hilarious video demonstrating the point.

14 http://www.infowars.com/climategate-peer-review-system-was-hijacked-by-warming-alarmists

15 http://wattsupwiththat.com/2011/06/25/bring-it-mr-wirth-a-challenge

which I explain below;

- *ocean 'acidification'*—another misnomer because the oceans are not becoming acidic and never will in the next thousand million years (at least);
- *rising sea levels;*
- *weather disasters:* more storms, fewer storms, more droughts, fewer droughts, more hot weather, more cold weather, more snow, less snow: you name it, it gets blamed on global warming;
- *extinctions or threatened extinctions,* most famously the case of polar bears;
- *'climate refugees':* entire nations and peoples, it is alleged, will soon be fleeing lands that become uninhabitable due to warming[16];
- *wipe-out of coral reefs.*

This is by no means an exhaustive list. In discussing scientific questions, it is normally the fair and courteous thing to give a full statement of all the claims of the theory with which one disagrees. That is absolutely impossible here. Hundreds upon hundreds of disasters, major and minor, are blamed upon global warming, both by scientists and by the media. Just finding them all would involve an open-ended amount of research and would use all our available space; but possibly incomplete lists exist on the internet, of which one of the best-known is Numberwatch's Warmlist.[17]

Are there really many hundreds of actual, genuine threats against us as a result of global warming? Inspection of the items on the Numberwatch list dispels this fear immediately. Here are just a few of them:

- *Planet explodes.* Yes, really! Since there is no accepted mechan-

---

16 Indeed, in 2005, 50 million climate refugees were predicted for no later than 2010! The actual number: zero. See http://www.theaustralian.com.au/national-affairs/world-still-waiting-for-50-million-climate-refugees-by-2010/story-fn59niix-1226042490227.

17 Example sites include: http://www.numberwatch.co.uk/warmlist.htm, http://blog.heritage.org/2009/11/17/global-warming-ate-my-homework-100-things-blamed-on-global-warming.

ism for any planet to explode, and since Earth has spent most of its existence far hotter than even the worst of the alarmist prognostications, and since our sister planet Venus has been baked continuously for billions of years at temperatures that melt many metals and yet has not exploded, this claim is beyond ridiculous.[18]

- *Al Qaeda and Taliban being helped by global warming.* This one comes from "eleven U.S. generals and admirals", who say that droughts help terrorist groups; they then go beyond their competence level and stay 'on message' by doing the (by now) politically required thing and gratuitously blame global warming for the droughts.[19]

- *The world rots faster as global warming fuels fungi.* This example comes from a popular science magazine, and is based on a study of fungi in Britain, which they then extrapolate to the entire world, apparently on the assumption that the whole planet is just like the British Isles. How about actually going to hotter places, such as Australia, and observing that they are not overrun with fungi?[20]

- *Fish get lost.* This is another peer-reviewed scientific study. Apparently more warmth and CO_2 will cause fish to develop wrongly and lose their way to their breeding grounds. It would seem the brilliant success of fish over the past 500 million years of mostly hotter and more CO_2-ish climate was simply overlooked.[21]

---

18 Tom J. Chalko. *No second chance? Can Earth explode as a result of Global Warming?* NU Journal of Discovery, Vol 3, May 2001.
Wesley J. Smith, senior fellow in bioethics at the Discovery Institute and consultant to the Center for Bioethics and Culture writes about this claim: "One doesn't have to be a scientist to know this is beyond nutty."
(http://www.discovery.org/a/5911) See http://nujournal.net/core.pdf.

19 http://newsbusters.org/blogs/noel-sheppard/2009/10/10/abc-al-qaeda-taliban-being-helped-global-warming.

20 http://www.livescience.com/environment/070405_fungus_fruiting.html

21 http://www.news.com.au/weird-true-freaky/warmer-waters-leave-fish-floundering/story-e6frflri-1111115736812

- *Human race faces 'oblivion' from climate change.* Humans not only survived, but flourished during, the climatic optimum after the end of the last ice age, which was several degrees hotter than today and when at the start humans had nothing but stone tools to help them. Indeed, as this period progressed, it was the very time that humans 'came out', creating agriculture, writing, developing the first great civilisations, exploring and settling every continent except Antarctica, ceasing to be just one more species amongst many and rising to planetary mastery; so this one is too absurd for comment, and is only noteworthy because it comes from none less than the U.N. Secretary-General, Ban Ki-moon.[22]

- A vast array of claims, from the plausible to the fantastic. For example, from Warmlist's list[23] of claims about global warming, here are just the claims for the letter "F": *fading fall foliage, famine, farmers benefit, farmers go under, farm output boost, farming soil decline, fashion disaster, fever, figurehead sacked, fir cone bonanza, fires fanned in Nepal, fish bigger, fish catches drop, fish downsize, fish deaf, fish feminised, fish get lost, fish head north, fish lopsided, fish shrinking, fish stocks at risk, fish stocks decline, five million illnesses, flesh eating disease, flies on Everest, flood patterns change, floods, floods of beaches and cities, flood of migrants, flood preparation for crisis, flora dispersed, Florida economic decline, flowers in peril, flowers wilt, flying squirrels move up, fog increase in San Francisco, fog decrease in San Francisco, food poisoning, food prices rise, food prices soar, food production increased, food security threat (SA), football team migration, forest decline, forest expansion, foundations threatened, foundations increase grants, frog with extra heads, frosts, frostbite, frost damage increased, fungi fruitful, fungi invasion, fungi rot the world.*

---

22 http://seattletimes.nwsource.com/html/nationworld/2004065445_webclimate11.html. Ban Ki-moon seems to have developed a habit of putting his foot in his mouth on this subject, recently blaming severe weather on climate change—a claim that even alarmist scientists won't buy.

23 http://www.numberwatch.co.uk/warmlist.htm

To be sure we haven't overlooked something of vital importance, we must investigate the major alarmist claims amongst those at the start of this section, but we need not, and cannot, look at every one of the hundreds of claims that have appeared in the media, just a few of which are listed above. But the mere existence of such a "sky is falling" amount of alarm surely prompts us to ask why a single chemical, carbon dioxide, the necessary ingredient for life on Earth, has generated such a huge number of disconnected, often absurd, often contradictory alarms about future disaster? And why is it that all sorts of people with no scientific competence, such as politicians, generals, and UN chiefs, feel compelled to weigh in with uninformed opinions on a scientific question? Something *really* odd is going on here, and I'll share my thoughts on that with you in Chapter 6.

Preliminary Thoughts for the Defence

The next three chapters will look at the key issues of global warming alarmism. But first I would like to share some preliminary thoughts about this world-wide attack upon plant food.

Firstly there is the question of deception. You may recall that the alarm was, until some years ago, all about "global warming"; and then suddenly your local TV news stopped calling it that and started saying "climate change". Why? The shocking answer is that research was done to find out what terminology would trick the most people!

An academic paper exists that researches the best way to present the topic so that the public at large will mistake evidence *against* global warming as evidence *for* it: and the key change highlighted to do this is, precisely, to stop calling it "global warming" and instead call it "climate change". An astonishing document prepared for the Tyndall Centre for Climate Change Research starts[24]:

> "This paper uses a dynamic simulation model to situate the role of variables representing environmental processes in the social construction of the issues of climate change and global warming."

---

24   Bray, Dennis and Shackley, Simon: *The Social Simulation of the Public Perception of Weather Events and their Effect upon the Development of Belief in Anthropogenic Climate Change.* Tyndall Centre for Climate Change Research Working Paper No. 58 September 2004

I included this to give those readers who have been lucky enough to avoid this sort of thing an insight into how academics lost in a jargon-world can deceive so many, probably even themselves, with obscurantist verbiage. Roughly translated, that means "We made a model to figure out how using different names for environmental processes can influence what people come to believe about the subject." It shouldn't take the astute reader long to see what is being said between the lines. Moving on, what *did* they discover about the words that will influence people? Here's their words, where I have put the meaning I got from it in parentheses:

"We propose that in those countries where climate change has become the predominant popular term for the phenomenon, unseasonably cold temperatures, for example, are also interpreted to reflect climate change/global warming, and is indeed often reported as such by science through the general media."

(My understanding: Even though the scary theory is that we are all burning up, calling it "climate change" will trick a lot of people into thinking that cold weather is evidence that we are burning up.)

"In those countries where global warming has become the predominant popular term for the phenomenon of climate change/global warming, unseasonably cold weather is seen as a refutation of the phenomenon and indeed will lessen the belief temperature."

(My understanding: But if it is called "global warming", people won't make that mistake.)

"In effect, by constructing the model this way we have provided a means to address the impact of the way in which an issue is framed by science and the impact of the order in which the phenomenon unfolds in the physical world."

(This is close to jibberish, but I think it is saying: we are telling policymakers which choice will mislead the public, and which choice won't, so they have a means to pursue the policy they desire.)

And so now you know why this issue, which was called "global warming" for over a decade, suddenly got re-christened "climate change": it seems our political leaders have chosen to deceive us! All manner of weather can be blamed upon "climate change", including devastating cold. "It's so cold because of global warming" is obvious nonsense, but "It's so cold because of climate change" will fool quite a few people, even though the idea of global warming is actually what they are trying to sell us.

"Global warming" has been renamed precisely so that those who want to speak the truth cannot tell it to you. If a climate realist who knows the climate is always changing says "Yes, I believe in climate change", he will be misinterpreted as agreeing with the alarmist view that the climate now is changing for the hotter in an unprecedented way due to human interference through industrial emission of carbon dioxide. And if he says "I don't believe in it", he will be derided for denying the obvious fact that climate indeed changes. How is anyone supposed to even discuss this issue when our ability to talk about climate has been sabotaged?

I suggest that whenever an alarmist uses the phrase 'climate change', we should call them on it immediately, explain why this phrase was chosen to trick people, and ask them to please stop repeating this dishonest phrase. Depending on their response, you will then know the truth about their honesty or lack of it.

This is not the only dishonest term bandied about by the alarmists: anyone who disbelieves, or even entertains healthy scientific scepticism, is labelled a 'climate change denier'—a truly contemptible term devised to bring to mind the holocaust deniers and tar the scientific sceptics (i.e. the true scientists) with a neo-Nazi brush.

To put the record straight, the sceptics who retain scientific objectivity and those who, convinced by the evidence that there is nothing in the AGW theory, reject it outright, are actually climate change *realists*—they are the very people who have been saying all along that the climate *always* changes. It was hotter in the Roman warm period and the medieval warm period. It was colder during the dark ages and the little ice age. And so on.

On the other hand, the alarmists who hurl this insult at their opponents are often the same people who brought you the erroneous 'hockey stick' graphs, which represented that the world's temperatures stayed virtually flat for one or two thousand years until suddenly it shot upwards in recent times. Who, then, are the ones denying that the climate changes?

Unfortunately this is not a civil debate between two groups in honest disagreement, as the dishonest term "climate change" and the scurrilously false insult "climate change denier" testify. The ringleaders of the global warming movement have deliberately lied about critical matters on which billions of lives depend.

I once, as perhaps you have, watched Al Gore's movie, An Inconvenient Truth. That was before I did my own research into global warming, and, as perhaps you did, I trusted that, whatever errors the movie might contain, at least I would not be lied to, and I took Al Gore at his word. And yet the movie presents a graph of the last four interglacial cycles (which we shall look at in detail in Chapter 7), showing temperature and CO_2 marching up and down in lock-step, and yet 'forgetting' to tell us that the temperature changed first and the CO_2 followed some 800 years later. This means that the lock-step graph is no evidence at all for Al Gore's conclusion (in fact, it is compelling evidence *against* the theory).

A list of 35 alleged errors in Al Gore's movie has been compiled, and the first nine of them have been ruled in a British court as being inconsistent with mainstream scientific opinion[25], the judge only stopping because lack of time prevented his looking at the remaining allegations. The nine errors addressed by the court are:
- Sea level "rising six metres"
- Pacific islands "drowning"
- Thermohaline circulation "stopping"
- CO_2 "driving temperature"
- Snows of Kilimanjaro "melting"
- Lake Chad "drying up"
- Hurricane Katrina "man made"
- Polar bears "dying"
- Coral reefs "bleaching"

The court found that *all* these claims are false. If the global warming theory is true, why is it necessary for its supporters to tell lies about it? Wouldn't the evidence speak for itself?

We shall see in the following chapters how the temperature re-

---

25 Christopher Monckton: 35 Inconvenient Truths: The errors in Al Gore's movie. http://scienceandpublicpolicy.org/monckton/goreerrors.html

cord has been falsified, and how vast amounts of government and other money has been made available exclusively for those who generate more 'evidence' for the AGW theory. Also, billions of dollars of business has now been built around flow-on activities such as providing 'green' energy and trading in carbon markets. All of this locks people in to supporting the theory. There would be no problem with expenditure if it was indeed in areas that are helping the planet, but in these forms, it is accelerating the harm being done.

The vast majority of people who believe in AGW do so because they have, quite naturally, trusted the 'scientific' authorities who have told them that the world is burning up due to human influences. Science in our time has been so badly corrupted by political influences that the real scientific method has effectively ceased to operate, but this fact is little-known except by those who have investigated the question in-depth. But as we each go about our busy daily lives we cannot investigate everything in depth, and this corruption is so unexpected that no one can possibly be blamed for wrongly assuming that science is still operating in the proper way.

Pushing on, then, past the semantic fog, climate realists understand that the climate has changed throughout the entire history of the world, all 4.5 billion years of it, and it is still changing. But with or without the dishonest terminology, the global warming theory, if true, would be a serious problem. We have three main questions to answer:

- Is global warming in fact bad?
- Is it in fact happening?
- Is CO_2 the cause of it?

All three of these questions must be answered in the affirmative for the anthropogenic global warming theory (AGW) to hold water. In other words, *all* the policies the alarmists are trying to sell us, such as carbon trading and emissions caps, are pointless or even damaging unless *all three of these questions* can be answered 'yes'[26]. So let us look at each of them.

---

[26] There is actually a fourth important question that also needs a 'yes' answer for there to be any policy issue for humans to 'solve': *Is the CO_2 increase due to human activity?* The answer to this question is probably 'yes', at least in part, but since the answers to at least two (the first and third) of the other three questions is a definite "no", this is something we need not be criticised for. Indeed, since CO_2 is beneficial, we should give ourselves a congratulatory pat on the back for its *positive* effects upon the environment!

Carbon is Life

2
Is Global Warming the Disaster They Say It Is?

The actions being proposed, and, as in Australia, enacted, to fight global warming have the potential to devastate the world's economies, with prohibitive power and fuel costs that will make even low to moderate energy usage impossible for many of the poorest, even in advanced western countries. Flow-on effects would raise the prices of everything else without exception. We shall look at this in more detail when we examine the precautionary principle in more detail, but I mention it now to highlight that it really is important that we understand the risks posed by global warming.

For one important example, James Lovelock of *Gaia* fame, had seriously claimed that "before this century is over billions of us will die and the few breeding pairs of people that survive will be in the Arctic where the climate remains tolerable."[27]. If that were likely, it might make turning off the heater and risking death from cold during winter a necessity. Lovelock's claims made news at the time (and maybe won more supporters to the alarmist cause) but recently he has understood how severely the alarmist hype is overstated—and more importantly, had the personal integrity to publicise his change of mind, saying: "The problem is we don't know what the climate is doing. We thought we knew 20 years ago. That led to some alarmist books – mine included – because it looked clear-cut, but it hasn't

---

27 James Lovelock, from *The Earth is about to catch a morbid fever that may last as long as 100,000 years*. http://www.independent.co.uk/opinion/commentators/ james-lovelock-the-earth-is-about-to-catch-a-morbid-fever-that-may-last-as-long-as-100000-years-523161.html. (Although Lovelock had apparently 'bought' the warming fallacy, it appears he has assessed other related issues correctly. This article also explains his thoughts on the value of humanity as the nervous system of Gaia, and of the damaging nature of wind farms and biofuels.)

happened."[28] We need to be aware, as Lovelock now is, that the science is far from "settled": if global warming is largely harmless or even beneficial, that would mean that it is not worth fighting, or perhaps that we should go so far as to welcome it.

Another reason we need to be clear-headed about the risks of global warming is that efforts expended fighting it will almost certainly subtract from efforts to fix other problems: we only have a certain amount of time, money, and effort that we can put into handling all the various issues facing us.

We must also remember that some problems require solutions that make other problems worse. For example, talking about the terrible problems of African poverty, respected African James Shikwati, the Kenyan economist and Director of the Inter Region Economic Network, argues that the real danger is being caught without modern technology, including electricity, to fight dangers such as famine and disease:

> "If you were to ask a rural person to define development, they'll tell you, yes, I'll know I've moved to the next level, when I have electricity. Actually not having electricity creates such a long chain of problems, because the first thing you miss is the light. So you get that they have to go to sleep earlier, because there's no light. There's no reason to stay awake. I mean, you can't talk to each other in darkness. ...
>
> The question would be how many people in Europe, how many people in United States are already using [renewable] energy? And how cheap is it? You see, if it's expensive for the Europeans, if it's expensive for the Americans, and we are talking about poor Africans, you know, it doesn't make sense. The rich countries can afford to engage in some luxurious experimentation with other forms of energy, but for us we are still at the stage of survival."[29]

Shikwati's critique takes on a whole new relevance for the west when we see the recent spectacle of politicians from both the left and the right calling having a reliable power distribution network "gold-plating"[30]. The idea is that peak power demand only occurs for short

---

28 http://wattsupwiththat.com/2012/04/23/breaking-james-lovelock-back-down-on-climate-alarm.

29 From Shitwati's interview in the documentary *The Great Global Warming Swindle*.

30 Australia's leftist prime minister and two rightist state premiers have jumped on

periods, and having reliable power during those times is an unafford-able luxury. As a child in the 1950s and '60s, I clearly recall having a completely reliable power supply in Brisbane Australia—apparently future generations will be expected to do without on a recurring basis. Power isn't always "just a luxury"; some need it for their very life support—such as the old and infirm in exceptionally hot and cold weather, the very times when the non-"gold-plated" system will be unable to cope. Instead of using our wealth to help lift the poor out of poverty, we are slowly regressing ourselves into third-world con-ditions.

Technology capable of lifting the grinding burden of doing things by human or animal power, while it can be supplemented by renew-ables, will largely require usage of fossil fuels, which will increase CO_2 emissions. Serious efforts have been proposed to stop the devel-opment of Africa in the name of fighting global warming. Shikwati adds:

> "One clear thing that emerges from the whole environmental de-bate is the point that there's somebody keen to kill the African dream. And the African dream is to develop. ... We are being told, 'Don't touch your resources. Don't touch your oil, don't touch your coal.' That is suicide."

The key point here is that all actions have consequences. It is in-credibly careless to react to an alarm (in this case, the scares about global warming) based on our own visceral fears, without thinking of the consequences for others perhaps far less well off than ourselves. This is horrendously unfair to the worlds least advantaged people, and if global warming isn't in fact a big problem, such behaviour is pointless as well, so it really is very important that we know, to the best of our ability, just how dangerous (or beneficial!) a warming planet would be.

So let us begin. As we have seen, the list of disasters that have been linked to global warming runs into the thousands, ranging from the deaths of polar bears and coral reefs to the planet exploding. It is claimed that it will cause droughts, floods, heat, cold, wind, lack of

---

this bandwagon. http://www.theaustralian.com.au/national-affairs/julia-gillards-power-blame-game-a-furphy-tony-abbott/story-fn59niix-1226446593729.

wind—you name it. But we shall see below that on virtually none of these issues is there the slightest evidence that global warming will indeed cause the threatened disaster—and on most issues we have good scientific evidence that it will *not*.

The Belief that Warmth Is Bad Contradicts Everything We Know from History

Two thousand years of historical records also tell us unambiguously that the weather is far, far worse during cold times than warm ones. Ours is the *first generation ever* of human beings who fear warmth. Every previous generation knew that the killer was the cold—almost every culture has a 'Christmas': a mid-winter feast to celebrate having survived to the deepest point of the yearly cold and to welcome the return of Spring and warmer weather. The dark ages commenced with the onset of cruelly cold weather. Frosts wreaked havoc with agriculture, diseases spread—the bubonic plague killed millions.

Eventually the warmth returned, and in the medieval climate optimum (note in passing how our predecessors named the period of warmth!) the now-frigid Greenland was settled by the Vikings, Europe prospered once again as it had during the Roman warm period, and the era of great medieval cathedral building flourished. If you are wondering how today's temperatures match up, melting glaciers in Europe have uncovered such finds as old Roman mines (proving that the glacier was missing in Roman times), whilst in Greenland, many of the abandoned Viking settlements are still under permafrost (again proving it was warmer back then—you can't plant a crop in permafrost).[31]

Our planet has one entire continent (Antarctica) and most of one huge island (Greenland) as well as major sections of other continents that are all more or less useless to life (excepting a very few specially-adapted species) due to ice cover and unremitting cold.

You might have seen news stories about people dying during heat

---

[31] And it seems Greenland ice is still growing. See some photographic evidence here: http://wattsupwiththat.com/2008/12/30/the-ice-in-greenland-is-growing.

Carbon is Life

waves (inevitably accompanied by the remark that this will become more common due to 'climate change'). But even this criticism of warmth fails to pass the common sense test. People die during cold spells too. The key question is not whether people die during warm spells, because we can't prevent either warm or cold spells. The question that matters is: which is safer? And remarkably, warmth proves to be far, far safer than cold.

A peer-reviewed study by Matthew Falagas and others examines excess deaths (over and above the natural death rate) during both warm and cold spells[32]. Their diagram summarising their country-by-country and month-by-month results clearly shows, for almost every country studied, a *declining* death rate all the way to the end of summer and an *increasing* death rate all the way to the end of winter. This means that, although both cold and warm spells kill people, the longer a cold spell lasts, the more the death rate ramps up: cold 'saps one's resistance'. But the longer a warm spell lasts, the fewer people die: we become accustomed to warmth, but not to cold. This pattern holds even in Australia, which is for the most part a hot country. Yet even here, a full 25% more people die at the height of the winter die-off than in the hottest month of summer!

Furthermore, warmth tends to kill those who are already weak, whereas cold can and does kill the young and healthy as well, so cold spells take away potentially more years of life per death (quite apart from the numbers of deaths) than warm spells do[33]. All across Europe, the Americas, everywhere the matter has been tested, cold kills more people than heat and steals more 'life time'.

---

[32] Matthew E. Falagas, *et al. Seasonality of mortality: the September phenomenon in Mediterranean countries.* Reported in a larger story at http://wattsupwiththat.com/2010/01/06/winter-kills-excess-deaths-in-the-winter-months.

[33] As reported in a BBC article quoting Dr Gavin Donaldson, a specialist in respiratory medicine at University College London. http://news.bbc.co.uk/2/hi/health/8442413.stm.

Why Should Today's Temperature Be the Perfect One?

The assumption that today's climate is perfect and we have to keep it the same at all costs seems to be taken for granted by the global warming alarmists, the mainstream media, and most politicians. But why on Earth should it be?

If anyone tells you "Because this is the temperature humans have evolved with", they don't know their history. The human species evolved over a period that covered the previous interglacial (the Eemian, around 125,000 years ago—warmer, more prolific than today, with sea levels four to six meters higher). This was followed by the last ice age (during which humanity suffered a crisis that almost sent us extinct, leaving only around 10,000 human beings on the planet). Then we entered the current warm interglacial, the Holocene, in which life flourished once more and human technological evolution accelerated. But the Holocene has already peaked and is now well past its maximum temperatures, with life in retreat from the colder zones, as always when the deadly cold returns.

It is often trumpeted that 1934 (in the U.S.) or 1998 or 2010 (worldwide) was the warmest year ever. But Don J. Easterbrook shows the evidence that:

> "Of the past 10,500 years, 9,100 were warmer than 1934/ 1998/ 2010."
> [34]

That fact alone discredits all the high temperature disaster alarms. The stark implication of 9,100 of the past 10,500 years being warmer than the warmest any of us alive today have ever seen is brought home in a different way by Richard Verney:

> "Perhaps the most graphic illustration of the benefits of living in a warm environment can be seen by comparing Stonehenge... with The Great Pyramid at Giza... . Both were built at approximately the same time, approximately 2,500BC, i.e., about 4,500 years ago.
>
> I would not like to devalue the enormity of the task behind the building of Stonehenge which was an absolutely fantastic achievement, but it pales in insignificance behind that involved in the build-

---

[34] http://wattsupwiththat.com/2010/12/28/2010%e2%80%94where-does-it-fit-in-the-warmest-year-list

ing of the Great Pyramid. Not merely upon the basis of the scale of each monument but also in the precision and building skills involved in the case of the Great Pyramid. In the latter, the base was chiselled out of the Giza Plateau to a horizontal accuracy of 21 mm over a length of 230 metres. We would be hard pressed to achieve such accuracy today. It must be remembered that this was achieved without the aid of self levelling liquids which we would use today. Stone blocks were chiselled and faced so that they could be set together with no more than a 0.5 mm gap between blocks. Compare this level of skill and craftsmanship with Stonehenge.

"The reason why such a difference of skill had developed was that in Britain, it was a struggle to stay alive in the relatively cold climate such that time was spent on staying alive rather than developing what were unnecessary and not relevant skills. On the other hand in Egypt, life was easy. The climate was generally benign and this allowed man to develop at a quicker rate than his counterparts living in colder European climes.

"Of course, I am not saying it is all down to warmer conditions. Available natural resources play their part but generally one can see a correlation between the date of development and warm environments. Historical evidence (of which there is plenty) would suggest that warmer conditions would be of significance to mankind."[35]

On a longer geological time scale, even the higher temperatures of the Holocene optimum are unremarkable—and on the low side. Over the 540-odd million years since the riotous explosion of multicellular life in the Cambrian period, temperatures have been several degrees colder than in the past ice age, and up to ten Celsius hotter than today (Figure 6[36]). For most of the time of life on Earth, our planet has been a hot, wet greenhouse—and life has flourished.

35 http://wattsupwiththat.com/2011/02/05/gavinology/#comment-592628
36 Temperature profile from http://www.scotese.com/climate.htm

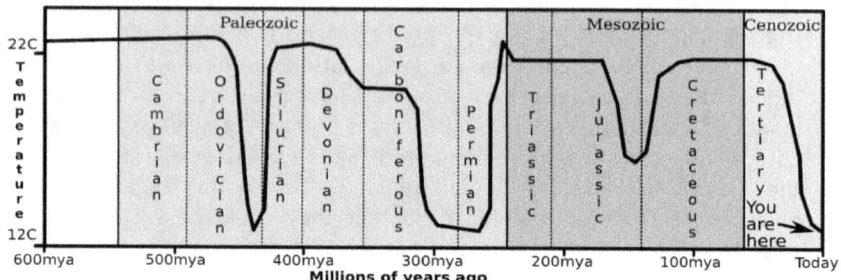

Figure 6: Earth's temperature is usually about 10C hotter than it is today

The frozen times and places have largely been times and places of death, not life. It should be obvious, but most of those alarming us about global warming never seem to mention it, that the Antarctic glaciers, barring a few highly specialised species, are devoid of higher life. But the Amazon rainforest isn't!

The deadliness of cold compared with the beneficence of warmth can seem counter-intuitive. One might think of a terribly hot day last summer and wonder how unliveable the Earth would be if it were 10C hotter than that, but it doesn't work that way. When the planet's average temperature goes up, it is mainly because the coldest places, where the ice restricts life, get warmer, thus allowing life to spread over more of the globe. Equatorial temperatures don't vary nearly as much. For example, if Antarctic temperatures go from minus sixty to minus twenty, the average global temperature would shoot up by quite a scary number, depending on exactly how you calculate it; but the facts on the ground would be that cold areas get more liveable and hot areas don't change much.

Also, the optimum temperature need not be a full 10C higher than today's for us to benefit greatly from a 3 or 4C warmer environment. The worst and most unlikely of the alarmist numbers is only 6.4C, which is a long way short of the 10C-hotter normal condition of our planet. Even if 10C hotter is a bit higher than optimal, the planet's history proves that life gets by just fine at even that temperature. So, far from being a disaster, the most likely outcome of planetary warming, on average, would be much, much better conditions for life. How strange that humanity fears something that has promoted life every time it has occurred in the history of our planet!

What about the Bad Effects of Global Warming?

Well, there are nowhere near as many of these, nor are they as serious, as the propaganda machine makes out, but also, it would be astounding if on an entire planet there were no ill effects whatever from a change in temperature—in either direction. The problem is we are being fed a steady diet of cherry-picked news stories about the bad stuff whilst the good stuff is studiously ignored (and most of the bad stories are untrue anyway). It is impossible here to answer every foolish thing that has been blamed on a warmer world (some of which were listed in the previous chapter). We have to restrict ourselves to some of the major, widely believed claims and see what is true and what is not. So, in no particular order:

Rising Sea Levels:

The idea here is that as the world's ice melts, the melt water flows into the oceans and raises the world's sea levels. The only ice we are concerned with is ice sitting on land. We don't have to consider the melting of the north polar ice or Antarctic sea ice any other floating ice, because when floating ice melts, it doesn't change the height of the water it is floating in[37]. (That is Archimedes' principle, for the discovery of which he reportedly yelled "Eureka!" and ran naked through the streets of Syracuse.) We only need to consider land ice, most of which is in just two places, Antarctica and Greenland.

Before we look at what is actually happening, let's ask what is the absolute worst thing that could possibly happen?

All the ice in both Antarctica and Greenland could melt.

What if it did? I calculated this for Antarctica. It is easy to do a

---

[37] This is not exactly true because icebergs tend to exclude salt as they freeze, so they are slightly less dense than they would be if they were unaltered frozen sea water. If all the sea ice melted it would raise sea levels by about 5mm You can see my calculations here:
http://peacelegacy.org/articles/does-melting-sea-ice-raise-sea-levels

'rough and ready' calculation as follows: Look up on the web the total Antarctic ice volume (it's about 30,000,000 cubic km). Now assume it all melts. So multiply that by the density of ice (about 0.91) to get the equivalent volume of water. Assume it is spread all over the oceans, so look up the area of the world's oceans and divide the water volume by that number. You should end up with a value of about 70 metres.

That figure is not accurate for all sorts of reasons (compression of ice and water changes the density, the rising water will cover a greater area as it inundates land, land will change its height due to a phenomenon called isostatic rebound, and so on), but there is something really nice about actually calculating a number for yourself and seeing with your own eyes the reason for the number. A presumably more accurate figure from NASA[38] is 60 metres—not too far off the rough and ready number. The melting of Greenland's ice would add another 6 metres, giving about 66 metres of sea level rise if *all* the ice in the world melted.

So the very worst? 66 metres sea level rise. How long would that take? The same NASA announcement tells us today's melt rate is 100 cubic km per year, so from the total ice volume (see above) divided by this number, we see that at the present rate, it would take 300,000 years to melt all that ice and get that 66 metre sea rise. If the melt rate were *fifty* times today's, which would require scorching temperatures that no one is predicting, the melt could happen in 6,000 years at a rate of ten millimetres sea level rise per year. That is the absolute worst that could possibly happen. And in the process we get a big island and a huge continent freed from ice for humans and other species to colonise.

So now let us turn to the real world. Yes, sea levels are going up—at the same rate (or lately, a bit *less*) ever since the end of the 'Little Ice Age' in the early nineteenth century. They have been rising at various rates ever since the end of the last major ice age, and today's rate is almost the smallest it has been in all that time. Satellite measurements show a flattening off, not a speeding up, in the twenty-first

---

[38] http://earthobservatory.nasa.gov/Newsroom/view.php?id=42399

century.

Even NASA, whose web pages are usually devoted to 'selling' the official global warming story, includes the graphic shown in Figure 7[39], showing that sea level rise recently is far less than it has been for most of the past 20,000 years.

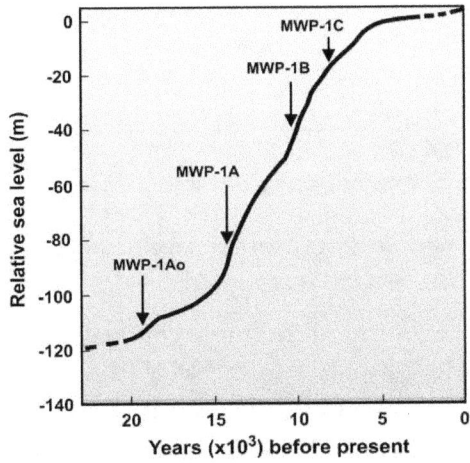

Figure 7: NASA's own graphic shows long-term sea level rise

The little uptick in sea level rise at the very end of Figure 7 has recently stopped. A recent peer-reviewed paper[40] estimates that sea level rise attributable to ice mass loss is only 0.54 mm/yr, or about two inches per century.

The University of Colorado at Boulder supplies sea level data. Their home page at the time of writing shows the graphic in Figure 8[41], showing a flattening since about 2005. This is a comparatively short time and one cannot rely on the slow-down being permanent (or even real), but at minimum it proves there is no evidence behind claims that sea level rise is accelerating.

---

39 http://www.giss.nasa.gov/research/briefs/gornitz_09/

40 Wu, Z., Heflin, M.B., Schotman, H., Vermeersen, B.L.A., Dong, D., Gross, R.S., Ivins, E.R., Moore, A.W., and Owen, S.E. 2010. Simultaneous estimation of global present-day water transport and glacial isostatic adjustment. *Nature Geoscience* **3**: 642-646.

41 http://sealevel.colorado.edu/index.php

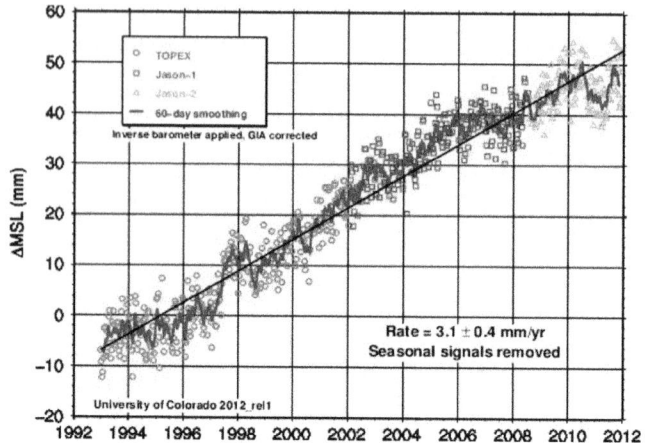

Figure 8: Sea level rise is not accelerating

They also have a 'wizard' that lets you make your own sea level plots since 1993 for any place in the world's oceans[42]. This is an interesting exercise because we can investigate for ourselves very easily. For example, I made the plot shown in Figure 9 for a spot in the Atlantic just off the U.S. coast.

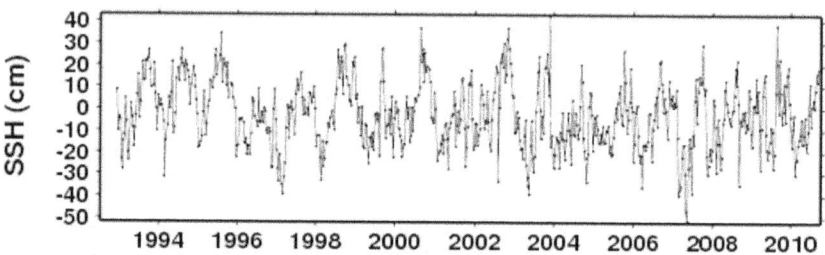

Figure 9: Sea levels at longitude 318, latitude 26

It is remarkable, seeing how much alarmism we get in the media, how very flat this diagram is. And it isn't a special or unusual spot—it's much the same anywhere, as you can verify for yourself.

Even the worst of these figures is a hundred times less speedy than some alarmists would have us believe, but on top of that, sea level rise is also nowhere near as damaging as we have been told. We are told Bangladesh will be flooded. Bangladesh occupies the mouth

---
[42] http://sealevel.colorado.edu/wizard.php

of the Ganges river, which is a huge delta of river channels and silt islands. The silt was brought down the river during floods and deposited where the swiftly flowing river ran into the sea and was forced to slow down. If the sea rises, the river will hit it sooner and drop its silt sooner and build up the level of the delta to the new higher sea level. The delta isn't at sea level by some astronomical coincidence, it is at sea level *because* it is the sea level. If the sea rises or falls, so will the height of the islands in the delta—they are not rock, but shifting silt. The system will adjust; we're not talking about some huge change overnight: the rate of sea level rise is less than the speed of your fingernails growing.

And to cap off the Bangladesh sea level fiasco, it appears the truth is that Bangladesh is actually *gaining* land from sediment deposition at the rate of about 20 square kilometres a year.[43] According to government official Maminul Haque Sarker[44], Bangladesh has gained about 1,000 sq km since 1973. His prediction, from one who, unlike the alarmists in the west, has to actually live in the country, is that they will probably gain another 1,000 sq km in the next 50 years.

A similar phenomenon is happening with the Pacific Ocean's coral atolls; the coral grows to suit the height of the ocean, whatever that may be, up or down. But the atolls have a lens of fresh water just beneath the land surface, on top of the salt water. The fresh water comes from rain and remains floating due to fresh water being less dense than salt water. But pump off too much of that fresh water for human uses, and the obvious happens: the land sinks. Some island communities are managing this problem well, others are not. The most disastrous response possible is to blame the problem on a false cause (global warming) so that the people take no action to counter the real cause. This is just one more case where telling fibs about science causes real harm to real people.

---

[43] http://rogerpielkejr.blogspot.com.au/2009/09/incoherence-continues-on-adapatation.html

[44] Deputy Director of the Center for Environmental and Geographic Information Services (CEGIS).

Acidification of the Oceans:

It is hard to believe just how bad the 'science' is behind claims that the oceans are turning acidic. It isn't hard to see why, but we need to look at some simple chemistry: namely, what it means to say something is "acidic".

Acidity is measured by a number called *pH*. This ranges from 14 down to zero. Values below 7 are acidic, 7 is the pH of pure water, and numbers above 7 are alkaline. (Chemists call alkaline substances *bases*.) Table 2 shows the pH values of some common substances. It will surprise many that even milk is officially "acid" because it has a pH of 6.6.

pH	Substance	
14.0	Sodium Hydroxide (caustic soda)	Base (alkaline)
12.4	Lime (calcium hydroxide)	
11.0	Ammonia	
10.5	Milk of Magnesia	
8.3	Baking Soda	
8.1	Sea water	
7.4	Human Blood	
7.0	Pure Water	Neutral
6.6	Milk	Acid
4.5	Tomatoes	
4.0	Wine & Beer	
3.0	Apples	
2.2	Vinegar	
2.0	Lemon Juice	
1.0	Battery Acid (contains sulphuric acid)	
0	Hydrochloric Acid	

Table 2: pH values of some common substances

So what does a chemical's pH mean? Water is the molecule H_2O, as most people are aware. In other words, it is two hydrogen atoms

(H) attached to a single oxygen atom (O). But in pure water, a small number of molecules 'dissociate': they split up into a positively-charged hydrogen *ion*[45] (written H⁺) and a negatively-charged hydroxide ion (written OH⁻). This happens to very, very few water molecules: about one in every 10,000,000 molecules splits up like this, which means that there is one H⁺ ion for every 10,000,000 water molecules. Now 10,000,000, in mathematical notation, is written 10^7 (the seventh power of 10). That "7" is the pH value: it is nothing more than the ratio of water molecules to hydrogen ions, written as a power of ten. This means that the *lower* the pH, the *more* hydrogen ions there are and the *more* acidic a substance will be.

For a few more examples, let's take some of the substances listed in Table 2 and see what this means:

- Caustic soda has pH 14: meaning there is one H⁺ ion for every 10^{14} (i.e. 100,000,000,000,000) water molecules;
- sea water has pH 8.1: there is one H⁺ ion for every $10^{8.1}$ (i.e. about 126,000,000) water molecules;
- apples have pH 3: meaning there is one H⁺ ion for every 10^3 (i.e. 1,000) water molecules;
- battery acid has pH 1: meaning there is one H⁺ ion for every 10^1 (i.e. 10) water molecules;
- hydrochloric acid has pH 0: meaning there is one H⁺ ion for every 10^0 (i.e. 1) water molecule.

There is an extremely important fact about the pH scale that can be seen in these examples: to decrease pH by *one*, you have to have *ten times* more H⁺ ions. A scale like this is called a logarithmic scale because multiplying the thing being measured only results in a change by a fixed amount to the value of the measurement (in the pH case, each multiplication by 10 decreases pH by 1). Another well-known logarithmic scale is the decibel scale for measuring sound, in which the number of decibels goes up by 3 for each doubling of the intensity of the sound wave. This odd value was chosen so that an extra ten decibels corresponds to a ten times more powerful sound—for example, a person talking a metre away is about 60 decibels; but just

---

[45] "Ion" is just a name for any charged atom or molecule.

10 more, or 70 decibels, is the loudness of a vacuum cleaner a metre away. Another well-known logarithmic scale is the Richter earthquake scale, in which each number in the Richter scale releases 30 times more energy than the number below (thus a Richter 6 earthquake might cause buildings to shake and plaster to crack, but a Richter 7 earthquake might cause bridges to twist and some buildings to collapse[46]).

The reason this is important for our story is that—as the decibel measures of a talking person and a vacuum cleaner illustrate—on a logarithmic scale you need *really big* changes in the thing being measured to make small changes in the number on your scale. Here's what that means for us: to get from pH 8.1 (the pH of sea water) to pH 7, the pH of crystal pure distilled water, you need *over ten times* the hydrogen ion concentration. A change by a mere one pH is a really big change.

'Acidification' means 'to be made acid'. We have seen what an immense change that would require in the hydrogen ion concentration. Can a change in the atmospheric CO_2 content achieve this? Well there is over 70 times the CO_2 in the oceans as in the atmosphere, so if we doubled the atmospheric CO_2 concentration and *all* of that extra CO_2 went into the oceans, it would make a change of 1 part in 70. This is laughably short of the factor of ten needed to get even close to the pH of pure water, let alone to any dangerous level of acidity. The oceans are alkaline and they are going to stay alkaline for a few billion years more. The oceans are not acid nor are they being made acid. This false claim is based entirely on misuse of language intended to mislead.

Steven Goddard writes[47] about this deceptive claim as follows:

"[Acidification] sounds very alarming, so being diligent researchers we should of course check the facts. The ocean currently has a pH of 8.1, which is alkaline not acid. In order to become acid, it would have to drop below 7.0. ... corals became common in the oceans during the Ordovician Era – nearly 500 million years ago – when atmospheric CO_2 levels were about 10X greater than they are today."

---

[46] http://www.matter.org.uk/schools/content/seismology/richterscale.html
[47] http://wattsupwiththat.com/2009/01/31/ocean-acidification-and-corals/

Claims of ocean acidification ignore the facts of our planet's history, in which atmospheric CO_2 has been twenty times today's levels and yet the oceans have not become acidic. And the oceans are huge, and contain over 250 times more matter than the entire atmosphere. CO_2 is less than one thousandth of the atmosphere. Even if it all dissolved in the oceans tomorrow, there is not enough CO_2 in the entire atmosphere to turn the oceans acid. And if humans burned all the fossil fuel in existence, it *still* wouldn't happen.

And to cap it all, recent research has shown that sea creatures forming calcium shells, contrary to expectations, actually do better, not worse, in a less alkaline ocean.[48]

The simple proof that CO_2 doesn't harm coral is the fossil evidence of vast coral reefs at geological times with far more heat and far higher CO_2 levels. We have proof by existence that coral thrives with lots of heat and CO_2.

Habitat Loss:

A moment's thought will show that if parts of the dead, frozen regions of the planet thaw, they can become productive and provide more habitat for both humans and other creatures. And we should remember that every single species on Earth today survived the Holocene climate optimum some eight thousand years ago, which was warmer than today, and they also survived the Eemian interglacial (before the latest ice age), which was even warmer than the Holocene.

This means that warming *increases* available habitat. However, we cannot relax too soon, because there is another way warming might cause trouble: even though there is more habitat than there was before, it is in different places; in other words, the climate zones shift.

We would expect climate zones to shift according to temperature changes, with warming shifting them towards the poles. Animals

---

[48] One estimate is a 13% *greater* calcification rate in the past fifty years. See Sherwood, Keith and Craig Idso, http://www.co2science.org/articles/V12/N22-/EDIT.php.

should be able to move with the shifting climates—after all, we are talking fractions of a degree per year, even for the scariest predictions; but how about plants? Evidence shows that as the Earth came out of the past ice age, plants that used to grow in southern USA, for example, now grow in Canada. Checking out the danger for every plant species would be a huge undertaking, but we do know that every plant on Earth today survived the very rapid warming as the ice age ended.

To see how rapid these temperature changes can be, I checked the historic rates recorded in the Vostok ice cores[49]. Between 211 and 190 years ago[50], temperatures rose by 1.31 C, or over 6C per century, or about the same as IPCC's maximum estimate (which, as we shall see in the following chapters, is greatly inflated). Contrariwise, between 397 and 375 years ago, temperatures fell by 1.1C, or about 5 degrees per century.

All plant and animal species alive today survived these changes.

Also, with temperature changes over the twentieth century not even a fifth of these values, it is clear that current changes are way below the changes that happen due to perfectly natural climate variation.

Finally, if there does turn out to be a plant that cannot propagate its seeds fast enough, or an animal blocked from migrating due to human disruption of habitat, wouldn't it be heaps easier to give that plant or animal some assistance? The "can't move fast enough" argument seems to assume that humans would simply sit by dumbly and watch a disaster unfold instead of lending the simplest kinds of help. I think there are enough friends of the natural world, including you and me, to prevent that.

Habitat loss is a real problem, but mainly due to humans' poor usage of land (breaking up environments, inappropriate dams, destruction by (real) pollution, and so on). These are real problems we should be doing something about, but they have nothing to do with global warming, so yet again, it would seem we are ignoring a real

---

[49] http://www.ncdc.noaa.gov/paleo/icecore/antarctica/vostok/vostok.html

[50] It is not clear from their notes just which year they consider to be "now".

problem (that is, letting it get worse) whilst worrying ourselves need-lessly about an illusory problem.

Most of the Warming Would Be in the Coldest Places

The climate models, which are the only reason to think that we are on a course to warming disaster, predict that the coldest places will warm more than the hotter ones, winter will warm more than summer, and minimum temperatures will increase more than maxima. This is also the understanding of how climate changes for natural reasons. So according to the entire range of scientific opinion, including alarmists and realists, the high-latitude polar regions will warm the most; this is because the strongest infra-red emission from polar ice lies in or near the strongest absorption wavelengths of carbon dioxide, which magnifies the effect (according to the climate models) near ground level at the poles. In other words, frigid places will become habitable but hot places won't get much hotter.

Suppose temperature in Antarctica increased by 15C: That might take the place from -60C to -45C. It's still nowhere near thawing, but the world's average temperature might go up by 1C as a result. In short, world average temperature doesn't mean much unless you know where and how the changes are distributed. The way a planetary average, or *mean*, temperature, is computed is largely arbitrary, by combining temperatures measured over many places on the Earth's surface. Many scientists even argue that the planet has no meaningful "average" temperature, but we need not agree to that to see that a lot of warming in really cold places can change the number in a scary way without any actual harm resulting.

In this way, even extreme warming (far more than predicted) would be beneficial (imagine northern Canada, Greenland, and Siberia covered by lush prairies, forests, and crops). The deadly cold areas would become much warmer (and therefore habitable) whilst the already hot areas would only become a little hotter. This means the total habitable area of the planet grows larger, most likely increasing the habitable zone for all or almost all species.

Another positive effect is that plants in many cold-limited places would have longer growing seasons, leading to more and better harvests and more food for wildlife.

Claims about Increased Storms and Hurricanes Are Baseless

Hurricane energy is flat or has even decreased, not increased, and has approached its lowest value in thirty years. In 2009 Eastern United States there were almost no hurricanes. Figure 10[51] shows the situation in the North Atlantic.

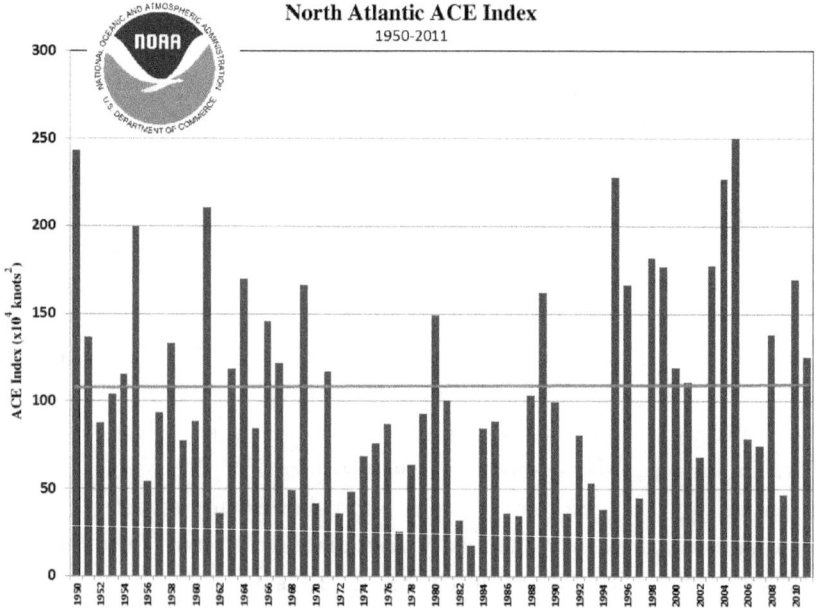

Figure 10: Hurricane energy has been flat for 60 years

The reason storms will not increase (and might decrease) is that extreme weather is caused by differences in temperature, pressure

---

51   http://www1.ncdc.noaa.gov/pub/data/cmb/images/hurricane/2009/annual/2009
     NorthAtlanticACE.png

etc., not by their absolute value. Storms unleash huge amounts of energy and they get it from differences in energy, for example, wind rushing from an intense high pressure area to an intense low pressure area. Tornadoes also show no cause for alarm, as shown in Figure 11[52].

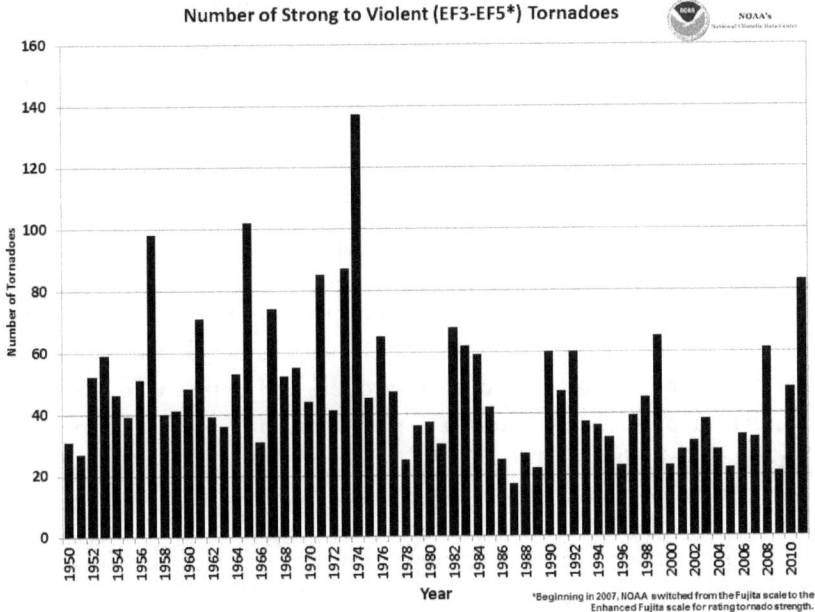

Figure 11: The number of strong to violent F3-F5 tornadoes is also flat

Historical evidence also shows that cold times have worse weather than warm times. The Little Ice Age and the Dark Age were times of intense, terrible storms. The Medieval Climatic Optimum (warmer than today) was a time of relatively balmy weather.

We can't leave this topic without a mention of this year's hot sky-is-falling topic, hurricane Sandy, which devastated parts of east-coast USA just before the presidential election, giving rise to a media frenzy of claims about global warming. Fact: extreme weather happens, regardless of the temperature, and Sandy doesn't change the fact that we currently have fewer, not more, extreme damaging

---

52 http://www1.ncdc.noaa.gov/pub/data/cmb/images/tornado/clim/EF3-EF5.png

events. Caleb Shaw compares Sandy with the great hurricane of 1821, convincingly showing that the 1821 event, as a hurricane, was much more severe than Sandy. Sandy's peak flood effect was 2.68 feet higher than the 1821 storm, but:

- Sandy hit at high tide, not low tide (6 feet difference);
- The full Moon added an extra foot to Sandy;
- Sea level in 1821 was a foot lower.[53]

On balance then, Sandy was 5.32-feet *less* of a storm than the 1821 event. If human-caused warming makes storms so bad, what caused the severity of the 1821 storm?

Claims about Droughts and Floods Are Baseless

This is a more subtle question than the simply wrong allegations about storms. Precisely where more or less rain will fall is currently not taken into account in climate models and is likely impossible to predict in principle. Professor Ian Plimer, one of Australia's most eminent geologists, writes:

"There are claims that human-induced global warming will increase variability in hydrological processes, especially floods, droughts and water supplies. However, droughts commonly occur in solar-driven periods of global cooling. ... The claims of catastrophes resulting from warming have not been supported by a demonstration that global warming will change the alternating wet/dry sequences, the period-icity of such sequences and the flood/drought severities. ...

"To Blithely state that global warming will lead to increased desertif-ication ignored previous validated science that shows that desertific-ation occurs during glaciation ... [The idea that modern climate change can lead to widespread and irreversible desertification] ... is contrary to all we know about the history of the planet."[54]

So yes, some areas would get more droughts or more floods if the world warmed up, and some would get fewer, but we don't know which areas are which, and as Plimer points out, *things would likely be even worse if the world gets colder!* And since predicting exactly where these influences will occur is likely not possible, it also means

---

[53] http://wattsupwiththat.com/2012/11/02/a-reply-to-hurricane-sandy-alarmists/
[54] Ian Plimer: Heaven and Earth, pp 136, 204.

that no one can know whether a particular drought or a particular flood happened due to global warming.

Predictions aside, real-world data to date shows no basis for the alarming claims in the media, where it seems every terrible disaster imaginable is going to start "tomorrow". But as for today, Figure 12[55] shows 600 years of drought reconstructions—just the same old patterns repeating over and over.

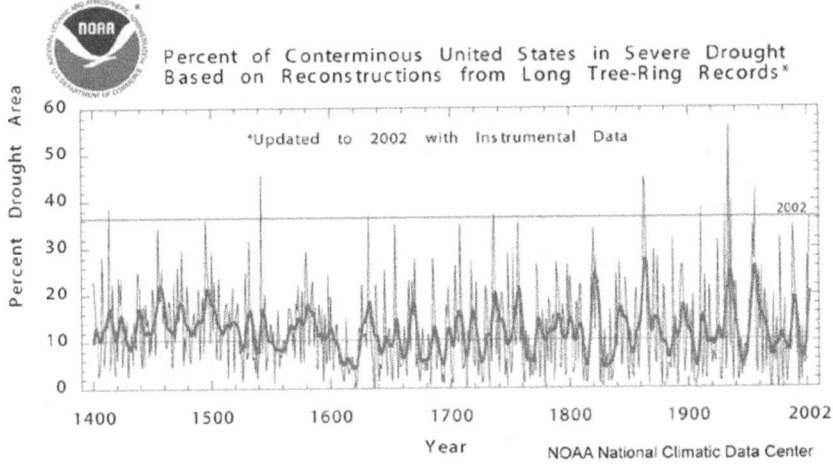

Figure 12: Long-term drought records show no worrying trends.

Likewise, the wet and dry records in Figure 13[56] taken from modern instruments show no significant trends at all. Shouldn't a really terrible looming disaster be a lot easier to spot in the data than this?

---

55 http://lwf.ncdc.noaa.gov/oa/climate/research/2002/ann/paleo-drought.html
56 http://www.ncdc.noaa.gov/sotc/drought/2011/13

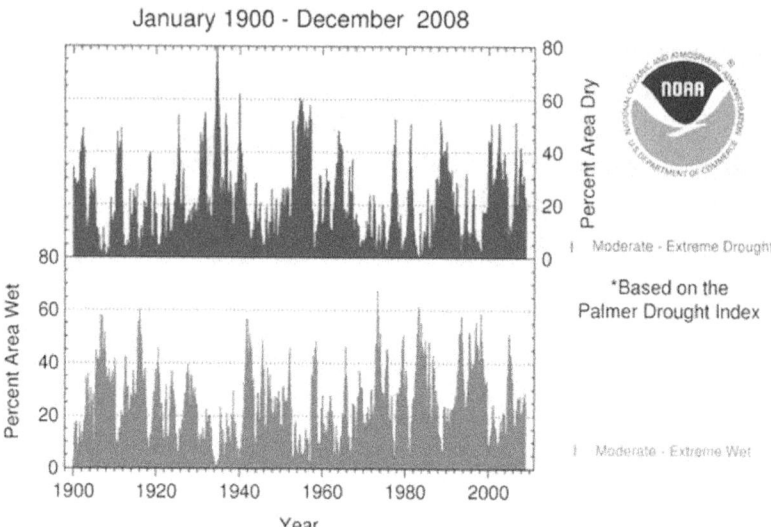

Figure 13: U.S. percentages of areas wet and dry since 1900

It does seem that the U.S. as a whole is getting slightly more rain, but since the continent's aquifers have been draining due to water use in agriculture, that would be a good thing, surely? See Figure 14.

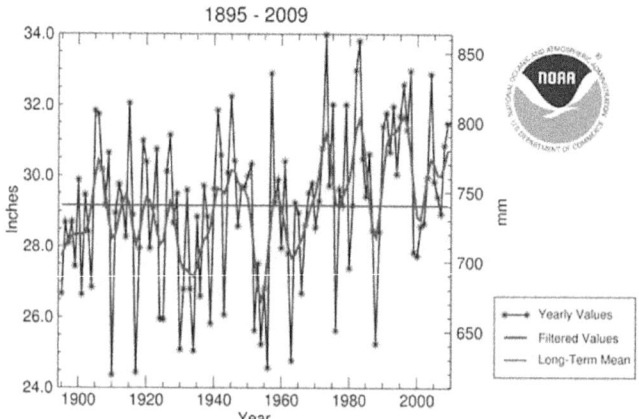

Figure 14: Precipitation in the contiguous United States since 1895 – a small but welcome upward trend

Historical experience proves that warming make all manner of trouble easier to handle and gives us, for example, better crops and

Carbon is Life

better places for wildlife.

Claims about More Diseases such as Malaria Are Baseless

Diseases such as malaria are diseases of poverty, not temperature. Malaria was endemic in parts of England from the fifteenth century onwards until the beginning of the twentieth century, and may have been there in Roman times. It was even endemic in Siberia! Professor Paul Reiter, one of the world's leading experts on infectious diseases, wrote on malaria in the "Little Ice Age" in a memorandum to the UK House of Lords[57]:

> "I wonder how many of your Lordships are aware of the historical significance of the Palace of Westminster? I refer to the history of malaria, not the evolution of government. Are you aware that the entire area now occupied by the Houses of Parliament was once a notoriously malarious swamp? And that until the beginning of the 20th century, "ague" (the original English word for malaria) was a cause of high morbidity and mortality in parts of the British Isles, particularly in tidal marshes such as those at Westminster? And that George Washington followed British Parliamentary precedent by also siting his government buildings in a malarious swamp! I mention this to dispel any misconception you may have that malaria is a "tropical" disease. ...
>
> "Despite this remarkably cold period, perhaps the coldest since the last major Ice Age, malaria was ... a "serious public health problem" in many parts of the British Isles, and was endemic, sometimes common throughout Europe as far north as the Baltic and northern Russia. It began to disappear from many regions of Europe, Canada and the United States as a result of multiple changes in agriculture and lifestyle that affected the breeding of the mosquito and its contact with people, but it persisted in less developed regions until the mid 20th century. In fact, the most catastrophic epidemic on record any-

---

57 http://www.publications.parliament.uk/pa/ld200506/ldselect/ldeconaf/12/12we2 1.htm
See also *From Shakespeare to Defoe: Malaria in England in the Little Ice Age*:
http://www.cdc.gov/ncidod/eid/vol6no1/reiter.htm.
And: *Global warming and malaria: knowing the horse before hitching the cart*:
http://www.malariajournal.com/content/7/S1/S3.

where in the world occurred in the Soviet Union in the 1920s, with a peak incidence of 13 million cases per year, and 600,000 deaths. Transmission was high in many parts of Siberia, and there were 30,000 cases and 10,000 deaths due to falciparum infection (the most deadly malaria parasite) in Archangel, close to the Arctic circle. Malaria persisted in many parts of Europe until the advent of DDT."

Polar Bears Are Not Drowning

The media-hyped polar bear panic has to be one of the most cynical campaigns of our age. Political activists and others, playing upon our natural desire to help these beautiful animals, have misused our concern for wildlife in the most disgraceful way that actually puts the bears in more danger, not less. More about that in a moment, but first let's look at the facts.

Polar bear numbers are not decreasing. A new study from the government of Canada's Nunavut province and reported in April 2012, showing that the polar bear population is stable, and some 66% higher than figures reported earlier.[58]

As for the effects of rising temperatures, H. Sterling Burnett has investigated this from the WWF's own literature—the WWF being a well-known 'pusher' of the global warming meme. He writes that polar bear populations are declining in regions (like Baffin Bay) that have had a decrease in air temperature, and they are increasing (near the Bering Strait and the Chukchi Sea) where there are increasing air temperatures[59]. This is the opposite of what we would expect if warmth is harming the bears.

You may have seen that photo of bears "stranded" on an iceberg and been told about bears having to swim so far they drown: the photo actually showed bears relaxing a few hundred metres from the

---

[58] Study at: http://env.gov.nu.ca/sites/default/files/ foxe_basin_polar_bears_2012-.pdf
Reported in: http://www.theglobeandmail.com/news/national/healthy-polar-bear-count-confounds-doomsayers/article2392523
A relevant article: http://wattsupwiththat.com/2012/04/05/nunavut-government-study-the-polar-bear-population-is-not-in-crisis-as-people-believed
[59] http://www.ncpa.org/pub/ba551

shore—they apparently climbed onto the iceberg to relax, or even to get a look at the passing boat from which the photo was taken. There have been zero known cases of polar bears drowning from global warming. Only four cases are known to have occurred as the result of a storm.

Dr Bjorn Lomborg, whose Copenhagen Consensus Center was ranked as one of the world's "Top 25 Environmental Think Tanks" by The University of Pennsylvania[60], writes that of the 20 populations of polar bears, two are growing, and only two are declining. "The habitats of the two thriving groups have actually become warmer."[61]

We do not, however, need to formulate theories about bears' swimming abilities to find out if they can survive hot weather. We are at the tail end of the Holocene interglacial, and as we have seen, it has been much warmer for most of the past 8,000 years than it is today; almost certainly, the entire Arctic has been free of summer sea ice a great many times in these 8,000 years. And the Eemian interglacial 100,000 years ago, the one before the last ice age, was warmer again, and polar bears survived right through all these hot times. As long as we do not mess up their environment in other ways, there is every reason to believe they can survive warmer weather now. But as long as we falsely delude ourselves that a bit of warmth is the big challenge for polar bears, we will not properly examine genuine problems they might face, from destruction of environment, to shipping, to human constructions in their living zones, and so on.

Capping off this entire fiasco is the fact that the WWF itself accepts that the main cause of polar bear deaths is lead poisoning—they are being shot.[62] The really criminal part of all this is that when faced

---

60 http://www.lomborg.com/about

61 http://www.telegraph.co.uk/earth/earthcomment/3310555/The-not-so-disappearing-polar-bear.html

62 The WWF writes: "Though much of the traditional harvesting from local communities has been sustainable, the IUCN Polar Bear Specialist Group (PBSG) documents that, both historically and currently, the main threat to polar bears is over-harvesting." From *Polar Bears at Risk* p14. awsassets.panda.org/downloads/polarbearsatrisk.pdf

with a chance to actually do something positive for polar bears, the very same government leaders who want to make you pay carbon taxes (allegedly to protect the bears) refused to do anything to help them. In March 2010, the Convention on International Trade in Endangered Species (CITES) voted down a proposal to ban polar bear hunting and trading in bear parts, and Canada still allows the hunting of 300 bears each year![63] The members of CITES are the very same governments whose leaders want to tax you crippling amounts in the name of saving the bears. Words fail to describe hypocrisy on this scale.

The whole polar bear scandal should tell us all just how much these 'compassionate' politicians who harangue us about global warming really care about polar bears or wildlife in general—or even about you and me! And it means there is a vastly easier way to protect polar bears than by wrecking our economies trying to make minuscule changes in the planet's temperature—ban shooting them! Global warming fear-mongering is a distraction from actually doing something useful for polar bears—and for other wildlife.

Claims of "Tipping Points" Don't Pass Reality Checks

A claim often heard is that, due to warming, we face a "tipping point" at which some bad disaster suddenly happens. A common version of this is that past a certain point, the temperature will run away uncontrollably. These claims face one insurmountable obstacle: as we have already seen, the Earth has been 10C or more hotter at various times in its history without activating any tipping points. It has had 25 times or more CO_2 without any tipping points. It has survived 4.5 billion years without runaway warming, despite being hit by doomsday-sized meteors, having supernovae explode nearby, and having supervolcanoes erupt so violently as to obliterate continents.

4.5 billion years is a huge time for a planet to survive in a danger-

---

63 National Resources Defense Council.
   http://www.nrdc.org/media/2010/100318.asp

ous universe without a heat apocalypse. Yet we are being told that the climate is so delicately balanced that adding extra CO_2 to the extent of 3% of the natural rate will suddenly activate a tipping point and runaway global warming will follow. The word "Venus" is usually mentioned in this connection, but in fact that just raises the question: Why isn't Earth *already* like Venus? And there is only one good answer to that question: whether we understand it yet or not, the planet must have a climate thermostat: natural feedback mechanisms must act so as to restore the planet to its normal temperature range. Earth isn't Venus, where the air is so thick (almost all CO_2) that an imaginary heatproof human being could swim in it! Venus has no liquid water, the critical substance for a thermostat because it can evaporate and form clouds. Thanks to liquid water, apart from some periods of deadly cold, the Earth has been temperature-stable.

Despite this proven track record on the part of our planet of successfully handling global warming, we are offered computer predictions from models that have already failed. And although there has been a small increase in temperature since the end of the Little Ice Age around 1850, Earth is well inside its known limits and nothing unusually bad has happened at all. And the planet has been turning green, as we shall see in a later chapter. In other words, we live on a planet that can feed hundreds of millions more than it could have done forty years ago.

Does this mean that there are no tipping points? No, but they won't be the ones the alarmists are scaring us with. Major changes in the planet's circumstances can cause the climate to change irreversibly. An example of this is the planet Mars, which once had a warm climate and liquid water, and perhaps even microbial life, but today it is dry and frigid, with no life or only the hardiest survivors hidden in crevices somewhere.

The reason Mars went downhill is that it was a bit too small: it lost internal heat rapidly (a word which, to geologists, means about 500 million years) and was unable to maintain a strong magnetic field like Earth's. So the atmosphere was at the mercy of the Sun's solar wind; ionised (electrically-charged) molecules were easily blown away into outer space, slowly removing most of the original thick atmosphere. Eventually a tipping point would have been reached at which liquid water could no longer exist; all the regulatory processes that operate on Earth to stabilise temperatures would have ceased, and the planet would have quickly frozen.

We can see in this example the hallmarks of a genuine tipping point: the planet's 'circumstances' slowly change until something is so materially different than it was before that the previous state cannot be maintained. In the case of Mars we are talking about the near-complete loss of its entire atmosphere. In the case of Earth and man-made CO_2, we are talking about changes measured in parts per million of a beneficial trace gas, well within limits that we know to have been safe when they happened before. There is no comparison between the genuine tipping point by which Mars lost its oceans and the 'sky is falling' scary fake tipping point of the climate alarmists.

When will a real tipping point happen in the case of the Earth? No one can say for sure, but here is one certain tipping point if nothing else gets us first: In about five billion years, the Sun will have used up all its hydrogen fuel and will start burning its helium; it will turn into a red giant star that will swell so much it actually engulfs the Earth. Then, if it hasn't happened earlier, the oceans will evaporate, the planetary thermostat will fail, and the Earth will heat uncontrollably and be burned to a cinder.

Any credible tipping point for our planet must involve one scenario: when the Earth uses up the last of its liquid water (that is, the oceans dry up); then it will be unable to regulate its temperature by producing clouds. That is a process that would take hundreds of millions of years. There is some debate about when or if this will occur before the end of everything in five billion years' time when the planet is swallowed by the giant red sun. But for sure it won't happen just because of CO_2 concentrations far less than those we have very easily lived with for hundreds of millions of years.

Expect Baseless Scare Stories to Keep on Coming

This book—or any other of the many excellent books now available that expose the global warming scare—cannot hope to answer every alarm you will see or hear in the mainstream media. The Australian Government alone has spent around 200 million dollars funding research into global warming. The Australian Coal Industry alone is spending a billion dollars funding 'clean energy' research, as we shall see in Chapter 5. And the U.S. and U.K. governments and the many European governments have spent unimaginable sums that dwarf this comparatively tiny expenditure by Australia.

In short, there is vast funding available for researchers who put up a case that represents that there is a danger to be investigated. So naturally, any reports or publications at the end of research funded by these grants have to justify the grant by producing a 'success story': "Yes, yes, we have indeed found *yet another* bad way in which CO_2 harms the planet!"

There are now many scientists who see through the scare, but there is precious little money available for them, which is why most are either self-employed, unemployed, fired, or retired and thus free from influence by their employer's agenda. A few tens of millions of dollars were once spent by an oil company funding sceptical research, and the story is endlessly reported in the mainstream media, whilst many *hundreds of thousands of percent more money* funding alarmist research is ignored, as if it has no influence on the science whatever.

The consequence of this is simple. No matter how thoroughly the scary claims about the essential plant nutrient CO_2 are debunked by largely unpaid climate realist researchers working in their spare time in between trying to earn a living doing something else, the lavishly-funded alarmists will always be able to produce more scares, and faster, than the realists can investigate and debunk them.

So expect the horror stories to keep right on coming. Another one I've heard is that narwhals in the Arctic are supposedly going to be drowned because the floating ice freed up by global warming will scoot around faster than they can swim and they won't be able to find any breathing holes (yes, really). This is at the same time as the polar bears are said to be drowning because they can't find any ice at all—I wonder just how gullible they think we are?

I wonder if the researchers who published this new scare took a moment to reflect that it was six degrees colder in the past ice age and two or three degrees warmer in the Holocene climatic optimum, and therefore both polar bears and narwhals have already survived the whole gamut of temperatures, with the Arctic either entirely ice-bound or entirely ice-free, and at some time or other with every possible state of iciness in between? The narwhal paper also demonstrates another feature typical of such alarmism: no apparent concern for a consistent story—because the narwhal authors reference the work of the polar bear authors, and yet the two alarms are apparently incompatible. Maybe if we fall for such idiocy we deserve whatever happens to us.

In short, even if you get this book the very day it comes hot off the presses, there will probably be some new and frightful alleged danger reported in the media that very day, which this book doesn't address. Keep watching web sites like the ones listed in the Resources section of this book, and the latest scares will soon be investigated.

Anyone clued in on the agenda behind this issue won't be alarmed by these scares or any new ones that can be reliably predicted to keep on turning up. And they definitely won't trust anything based solely on the word of politically financed 'scientists' who are committed to their conclusion at the outset by the source of their money. When a scare is 'on', most humans seem incapable of behaving rationally. The wise step back and wait, and check things out for themselves.

Finally, apart from the money, there are other reasons why CO_2 specifically is being attacked, which will be addressed in Chapter 6.

To Summarise...

We are fed a continuous diet of scare stories about a warmer planet; we are the first generation ever that is foolish enough to believe them—or even to think of them! Just as coral reefs survived and flourished through tens of millions of years at temperatures up to ten degrees Celsius warmer than today, and the vast herds of giant Jurassic herbivore dinosaurs ate tonnes of trees each, thanks to the wonderfully high level of carbon dioxide plant food in the atmosphere back then, and just as polar bears did perfectly well throughout the Eemian interglacial, two or more degrees warmer than now, just as the Roman Empire flourished and dug silver from land only now emerging from under glaciers, so, too, will we be better able to feed the groaning billions of human beings and also keep areas for wildlife, if we have a warmer planet with extended growing seasons and more cultivatable land, hopefully enhanced by an atmosphere richer in carbon dioxide plant nutrient than it is even now.

3

Even if Warming Is Bad, Is It Happening?

"It will without doubt have come to your Lordship's knowledge that a considerable change of climate, inexplicable at present to us, must have taken place in the Circumpolar Regions, by which the severity of the cold that has for centuries past enclosed the seas in the high northern latitudes in an impenetrable barrier of ice has been during the last two years, greatly abated.

"[This] affords ample proof that new sources of warmth have been opened and give us leave to hope that the Arctic Seas may at this time be more accessible than they have been for centuries past, and that discoveries may now be made in them not only interesting to the advancement of science but also to the future intercourse of mankind and the commerce of distant nations." - **President of the Royal Society, London, to the Admiralty, 20th November, 1817**[64]

As this quote from 1817 shows, changes in climate have been with us for a long time—and most of the trumpeted "unprecedented" changes of the past few years have indeed been precedented—and quite a long time ago at that! In chapter 2, we took a look at some of the huge body of evidence that, against everything we are told, a warmer world would most likely be a much better world. It always has been, and it almost certainly always will be. But is it even happening? Unfortunately, as we shall see, we cannot trust the official temperature records, but there are still ways to get at an answer to the question.

It is important to understand that there is not a single answer, yes or no, to whether the planet is warming:

- over the last 150 years it has warmed
- but over the last 1,000 years it has cooled
- and since 1998 it has been flat or cooling, not warming.

Currently temperatures are up a little on those of 150-odd years

---

[64] President of the Royal Society, Minutes of Council, Volume 8. pp.149-153, Royal Society, London. 20th November, 1817.

ago when the Thames froze over and times were tough (and it's a good thing too, and has saved about a billion people and countless wildlife from starvation). This is because of the inevitable warming since the end of the Little Ice Age around 1850. That's hardly surprising and it started before human industry ramped up and started emitting large amounts of CO_2. The warming has now reversed, despite increasing CO_2 emissions. But can we trust the precise numbers that the scary temperature graphs NASA, NOAA, Hadley[65], etc., all show us?

The Extent of the Warming Has Been Exaggerated

"Around 1990, NOAA began weeding out more than three-quarters of the climate measuring stations around the world. They may have been working under the auspices of the World Meteorological Organisation (WMO). It can be shown that they systematically and purposefully, country by country, removed higher-latitude, higher-altitude and rural locations, all of which had a tendency to be cooler."[66]

There was method in the madness of The National Oceanic and Atmospheric Administration (NOAA). By ignoring actual data from cooling or stable stations, their software fills in those areas of the planet with 'extrapolated' data from warming stations —supposedly 'nearby', but often hundreds of kilometres distant.

D'Aleo and Watts show us the graphic evidence of this selective dropping of measurement stations. Figure 15 is based on their presentation. It is easy to see that as the number of stations suffers a precipitous drop, the claimed temperature simultaneously undergoes a remarkable sudden rise. Was this an artefact in the data from dropping the stations, or was there a real cause? Do you remember any sudden huge change in 1991 that would explain a century's worth of temperature change? Nuclear war? Meteor strike? Supervolcano ex-

---

65 Hadley Centre for Climate Prediction and Research—a division of the UK Met Office.

66 Surface Temperature Records: policy driven deception? Joseph D'Aleo & Anthony Watts, Jan 29 2010, page 5.
http://scienceandpublicpolicy.org/originals/policy_driven_deception.html

plosion? I don't either; the simplest explanation (usually the best) is that the changes in the data introduced an entirely spurious increase in the temperature signal.

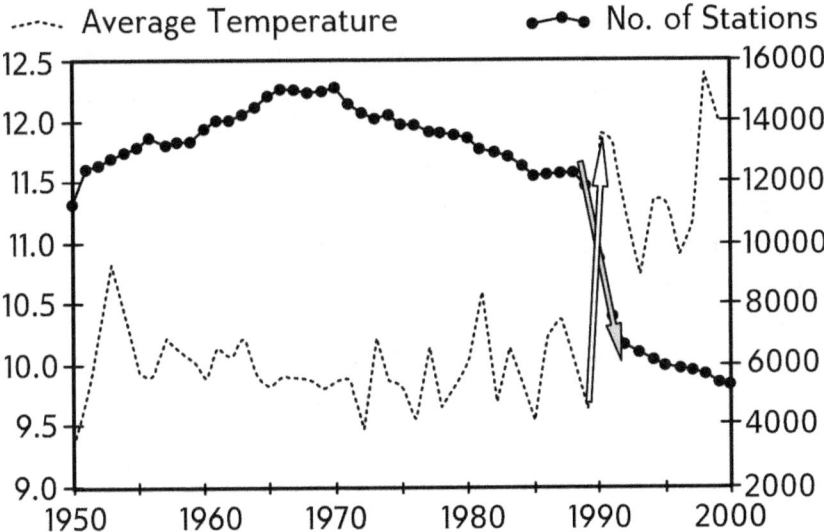

Figure 15: Number of temperature stations and the Earth's supposed sudden warming

The rise shown in this graph in 1991 accounts for the _entire amount_ of the 'global warming' that we are being told to wreck our economies to try to combat!

In their study of surface temperature records, D'Aleo and Watts conclude: "Instrumental temperature data for the pre-satellite era (1850-1980) have been so widely, systematically, and unidirectionally tampered with that it cannot be credibly asserted there has been any significant "global warming" in the 20th century."

What is happening here? Unfortunately our entire society is trusting manipulated data emerging from a very few places. As D'Aleo & Watts explain: "Five organizations publish global temperature data. Two – Remote Sensing Systems (RSS) and the University of Alabama at Huntsville (UAH) – are satellite datasets. The three terrestrial institutions – NOAA's National Climatic Data Center (NCDC), NASA's Goddard Institute for Space Studies (GISS), and the University of East Anglia's Climatic Research Unit (CRU) – all depend on data supplied by ground stations via NOAA." Many of those few key players have been compromised by scandal, including key members of Britain's CRU.

"Adjustments" to raw data are as big as the claimed warming; it seems that figures are adjusted retrospectively to show whatever current political imperatives require them to show.

Some examples of this have emerged recently in Australia. One of the best investigated is the temperature record at Darwin[67]. In brief, Darwin had five temperature stations operating at various times, telling a largely self-consistent story of slight cooling. In 'fixing' this data, the consistency amongst the stations was wrecked and unexplained 'adjustments' of over 2C (!) were added to the numbers to create a completely fictitious record showing warming. Figure 16, adapted from Wattsupwiththat's article, shows how this was done.

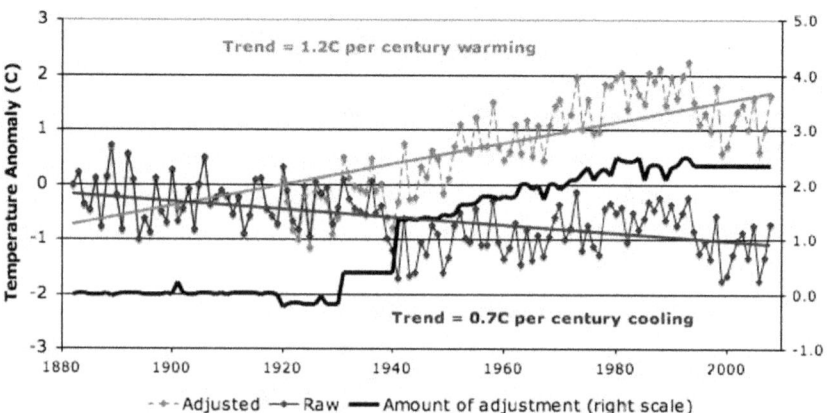

Figure 16: GHCN Raw and Adjusted Temperatures, Darwin Airport

The solid line in Figure 16 shows that entirely fictitious numbers were added to the actual temperature measurements made on real thermometers. The older values before 1940 were lowered by about 2C, whilst increasingly large numbers were added as the date approached the present. As more and more temperature stations come

---

67 The full story makes shocking reading; find it at:
    http://wattsupwiththat.com/2009/12/08/the-smoking-gun-at-darwin-zero

under similar close scrutiny, it is being found that Darwin's story is far from unique.

When I first decided to investigate this whole question, I felt I had to 'see with my own eyes', as it were. You can too. If you want to investigate for yourself, you can look up temperature data on your national meteorological databases. I did this for the stations close to me and, as have many others, I found that the actual raw numbers were far from alarming.

Another significant problem with the measurement record is bad siting or design of the measurement stations. The website surfacestations.org has done a major investigation of United States stations and found many reasons to suspect that poor siting is likely causing a spurious warming signal.[68] There is every mistake imaginable, from thermometers placed in the direct outlet path of air conditioners, to asphalt parking lots, airplane jet exhausts, a sewage treatment plant, fireplace chimneys, and more, making a mockery of any claim that these stations are recording temperatures that faithfully reflect the surrounding region.

Unfortunately these bad readings cannot be adjusted to discover the truth. If a temperature measurement is taken while a jet aircraft exhaust is pointed directly at the thermometer (it happens—airports are one of the commonest places to site temperature stations), then there is simply no way to correct that measurement to find out what the real temperature was: the record, sadly, is rubbish. Likewise, records taken at Arctic or Antarctic stations near the 'human' zone where snow and ice are cleared away, dark, heat-absorbing airport runways laid, and so on, have completely failed to do their job of recording for us a reasonably accurate temperature record. The numbers are baloney and can't be fixed. This is probably true of most of the temperature stations in the world.

What's more, the spotlight that has been shone on the shonky way the temperature records have been compiled hasn't made much difference to the approach of these leading climate research institu-

---

68   You can see with your own eyes at the photo gallery at:
      http://surfacestations.org/odd_sites.htm

tions. CRU have released a brand new version of their temperature record, Crutem4. All the figures in it have been 'adjusted', so it is fair enough, if we are being asked to live in energy poverty on the basis of these numbers, to ask CRU for the raw data. No such luck. According to CRU, you have to write to every Met office in every country on the planet and collect the data for yourself![69] In other words, you can't **check** their research: you have to **redo** their research—without the benefit of the millions of pounds of British taxpayers' money spent by CRU to make their original version. Good luck with that.

But the crazy part of all this is: even if all this manipulation of the temperature record were 100% accurate, current temperatures are still 'below average'! Currently the world's temperature is:

- colder than the medieval climate optimum and the Roman warming (which were times of plenty)
- colder than the previous interglacial, the Eemian, 100,000-odd years ago
- much, much colder than the Earth's geologically normal temperature range, the warm, wet hothouse, as Ian Plimer[70] calls it, in which life has existed for most of the past 600 million years (we are currently in a short respite in a 'deep freeze')

Again in 2012 the predicted outcry rose at the first sign of extreme heat in the early Australian summer, and the most overused word in the language, "unprecedented" was trotted out again. Global warming! But no one remembered the 1896 heatwave, so bad that special evacuation trains were run to get people out of the path of the heat![71]

---

[69] http://wattsupwiththat.com/2012/06/01/phil-jones-gives-a-talk-at-knmi-in-de-bilt-meanwhile-temperature-and-paleo-researchers-are-still-blowing-off-data-requests

[70] Ian Plimer. Heaven and Earth; global warming: the missing science. Connor Court Publishing, 2009.

[71] http://joannenova.com.au/2012/11/extreme-heat-in-1896-panic-stricken-people-fled-the-outback-on-special-trains-as-hundreds-die

The Hockey Stick Graph

If any one item has been most influential in convincing thousands that we face a "sky is falling!" heat catastrophe, it surely has to be the so-called hockey stick graph. It shows temperatures trundling along with little change for close to a thousand years, then—catastrophe!—they suddenly shoot upwards just when humans start running cars and trains and generally lifting masses out of grinding poverty. A picture shows a thousand words, so here it is[72]:

Departures in temperature in °C (from the 1961-1990 average)

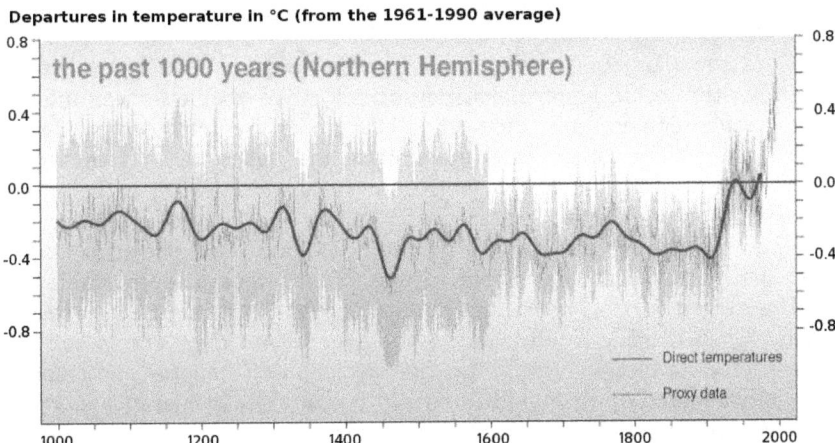

Figure 17: The Hockey Stick Graph

Figure 17 is extracted from a combined image that was included in the IPCC Third Assessment Report, 2001 (AR3). By the time the IPCC put out their Fourth Assessment Report (AR4) in 2007, the use of this graph had been greatly toned down.

Why did this graph have such an impact on the world's consciousness? Because for hundreds of years, it has been common knowledge that temperatures were warm and balmy in the medieval period, then cooled cruelly in the Little Ice Age, to finally rise again after about 1850. Historical records from many places and cultures, the records of cultures rising and falling in keeping with this understanding, and

---

72 From the IPCC Third Assessment Report 2001 Figure 2-3.
    http://www.grida.no/publications/other/ipcc_tar. Unfortunately no better URL can be given due to the site's use of html frames.

many proxy measures worldwide, all told the same story.

Yet—shock! Horror! The hockey stick graph showed we were all wrong for so long. Or so it seemed.

The full horror story of the rise and fall of the hockey stick empire would take an entire book to tell, and luckily Andrew Montford has risen to the challenge and done exactly that[73] in great depth. We cannot investigate here all the twists and turns in the full and disgraceful story of the hockey stick, and I recommend Montford's book to anyone who wants to hear the full evidence. But for the highlights? On his blog, Montford writes[74]:

> "At some time or another, most people will have seen the hockey stick—the iconic graph which purports to show that after centuries of stable temperatures, the second half of the twentieth century saw a sudden and unprecedented warming of the globe. This was caused, we were told, by mankind burning fossil fuels and releasing carbon dioxide into the atmosphere. For a while, the hockey stick was everywhere—unimpeachable evidence that mankind was damaging the planet—an impact that would require drastic measures to reverse. The stick's most famous outing however was just a couple of years ago when it made a headlining appearance in Al Gore's drama-documentary, An Inconvenient Truth. The revelation of the long, thin graph with its dramatic temperature rise in the last few decades, and the audience gasps that accompanied it, is something of a key moment for many environmentalists.
>
> "Shortly after its publication, the hockey stick and its main author, Michael Mann, came under attack from Steve McIntyre, a retired statistician from Canada. In a series of scientific papers and later on his blog, Climate Audit, McIntyre took issue with the novel statistical procedures used by the hockey stick's authors. He was able to demonstrate that the way they had extracted the temperature signal from the tree ring records was biased so as to choose hockey-stick shaped graphs in preference to other shapes, and criticised Mann for not publishing the cross validation R2, a statistical measure of how well the temperature reconstruction correlated with actual temperature records. He also showed that the appearance of the graph was due solely to the use of an estimate of historic temperatures based on tree rings from bristlecone pines, a species that was known to be problematic for this kind of reconstruction."

---

[73]   The Hockey Stick Illusion. A.W. Montford. Stacey International 2010.

[74]   http://www.bishop-hill.net/blog/2008/8/11/caspar-and-the-jesus-paper.html

I'll pass over the statistical technicalities entirely, but some questions struck me immediately I came across the hockey stick:

- First, this is just one proxy study; why should we—or even stronger, why *did* we—forget, seemingly overnight, hundreds of years of historical understanding about our planet, based on just one study—even if it had been done correctly?
- How did it get into the IPCC reports so fast, when our entire historical record tells against it?

The answers are enlightening but thoroughly demoralising. It turns out that there is a whole lot wrong with the science of the hockey stick, and worse, with the integrity of the process by which it was rushed into the IPCC Report. Montford continues:

"The controversy raged for several years, involving blue riband panels, innumerable blog postings, endless name-calling and dark insinuations about motivations and conflicts of interest. In May 2005, at the height of the controversy, and on the very day that McIntyre was making a rare public appearance in Washington to discuss his findings, two Mann associates, Caspar Amman and Eugene Wahl, issued a press release in which they claimed that they had submitted two manuscripts for publication, which together showed that they had replicated the hockey stick exactly, confirmed its statistical underpinnings and demonstrated that McIntyre's criticisms were baseless. This was trumpeted as independent confirmation of the hockey stick. A few eyebrows were raised at the dubious practice of using a press release to announce scientific findings. Some also noted that on the rare occasions that this kind of announcement is made, it tends to be about papers that have been published, or at least accepted for publication. To make such a dramatic announcement about the submission of a paper was unusual in the extreme."

Why such scientifically odd behaviour? Montford explains how the Wahl and Amman papers fell apart under the statistical analysis by McIntyre. But:

"As 2005 neared its end, two important events loomed large. The first was the year end deadline for submission of papers for the IPCC's Fourth Assessment Report on the state of the climate, and realisation soon dawned on McIntyre and the observers of the goings-on at GRL: the IPCC *needed* to have the Wahl and Amman papers in the report so that they could continue to use the hockey stick, with its frightening and unprecedented uptick in temperatures. Mountains were going to be moved to keep the papers in play."

And so it was. The plot is so Machiavellian and twisted that I don't think it can be summarised any better than Montford has done on his blog post. Critical statistics that blew apart the credibility of the paper were not included in the version that went into the IPCC Report. As far as the scheming and plotting side of this story goes, Montford summarises it:

> "The IPCC got their rebuttal of McIntyre and the journal got a fig leaf of respectability to cover up its duplicity."

All very well if you like a good soap opera, but what about the temperatures? Is the current climate unprecedented? The key scientific point to emerge from the grisly story sketched out above is that McIntyre showed that a key statistic for the significance of the analysis[75] was close to zero, meaning that it had no statistical significance. Climate Audit[76] is the place to go for McIntyre's expert statistical rebuttal of the hockey stick. Unfortunately, like the human interest part of this story, it is too complex to analyse here. The short summary is that the statistical method used in the hockey stick tends to 'find' hockey sticks everywhere. On his blog, McIntyre writes[77]:

> "You can get a hockey stick shaped PC1[78] from series in which there is no hockey stick shape in the underlying data. His method will pick out and overweight series with 20th century trends and flip the series so that the trends are all in the same direction. In the shaft of the stick as you get away from the common 20th century feature, the noise cancels out. Since the PC1 is a weighted average of the various series, the noise features cancel out. The variance of the average is small in the shaft, but large in the blade.

> "In the empirical situation of the North American tree ring data set about which there's been so much dispute, there actually are some hockey stick shaped series in the data set (bristlecones). Given that the "Artificial Hockey Stick" effect exists with red noise, it doesn't take much imagination to contemplate that it's really enhanced by actual hockey stick shaped series. But of you take out the bristlecones, there's no hockey stick (and bristlecones were known beforehand to be problematic)."

---

[75] A value known as R^2, the coefficient of determination.

[76] http://climateaudit.org. Ross McKitrick also has a nice overview page at http://www.uoguelph.ca/ ˜ rmckitri/research/trc.html.

[77] http://climateaudit.org/2006/03/27/new-scientist-on-the-hockey-stick

[78] "PC" stands for "principal component".

In fact the method used to make the hockey stick is so bad it flips proxies over so that cold means hot and hot means cold! McIntyre adds:

> "In our E&E article, we pointed out that if you artificially increased all the non-bristlecone tree ring proxies by 0.5 in the North American network for the period 1400-1450, the Mann method would flip all these series over and actually REDUCE the temperature index for the 1400-1450 period. It's a laughable method.
>
> "Now in the actual North American network, the hockey stick shaped series (bristlecones) all happen to be upward pointing hockey sticks, but the PC method does not use that information. They could be equally distributed half and half down and you'd get the same PC1. That's one reason why I don't like PC methods for these sort of systems."

There is a lot more to be told about this travesty of science and proper scientific processes, but no space to do here what Montford's book has already done. A last word on this subject from Rand Simberg[79]:

> "What does this all mean? First, let's state what it doesn't mean. It doesn't mean that we know that the planet isn't warming, and it doesn't mean that if it is, that we can be sure that it is not due to human activity.
>
> "But at a minimum it should be the final blow to the hockey stick, and perhaps to the very notion that bristlecone pines and larches are accurate thermometers. It should also be a final blow to the credibility of many of the leading lights of climate "science," but based on history, it probably won't be, at least among the political class. What it really should be is the beginning of the major housecleaning necessary if the field is to have any scientific credibility, but that may have to await a general reformation of academia itself."

Weather Is Not Climate!

The confounding of weather and climate is another big source of confusion. Climate scientists will tell you weather is not climate, and they're quite right. (But it's a pity they only seem to say it when the weather is cold.) What this means is that a particularly hot, or partic-

---

[79] http://pjmedia.com/blog/the-death-of-the-hockey-stick. Simberg also writes at http://www.transterrestrial.com.

ularly cold, or wet, or dry, or stormy, or balmy, day, month, or season does not mean, in itself, that the climate is changing. Climate is the 'summary' of the weather over extended periods of decades or longer. We have to wait and see whether a cold season presages a new ice age, or a hot one presages the global warming disaster we are continually being harangued about.

Unfortunately, the mainstream media and the 'climate scientists' pushing the alarmism don't take their own advice when the planet warms, or a drought or a flood or a hurricane happens. When any of these occur, you will probably see a mention of global warming (or its dishonest alias, 'climate change'). "Greenland glaciers are melting —climate change!" or "New Orleans hit by a hurricane—climate change!" or, strangely, even "Europe snowed in—climate change!"

So, even though isolated events, or even trends over just a few years, don't actually tell us much about 'climate change', nevertheless false claims are routinely made and then blamed on global warming.

Melting polar ice caps are a case in point. But Antarctic sea ice is growing, not shrinking. Arctic ice is as much or more affected by wind and sea currents as by temperature: the much-trumpeted and headlined 2007 summer melt was caused by winds blowing ice into warmer waters where it melted. The similar 2012 event happened when a huge storm smashed up the ice pack and drove the pieces into warm southern water.

Another example is shrinking glaciers: this is misleading and deceptive. Some are shrinking but many are growing. Glaciers are affected dramatically by changing humidity and rainfall. Mt Kilimanjaro is a case in point. Used as a talking point for global warming, the prime cause of its loss of ice was changing use of the surrounding land and loss of forest cover; and most of the loss had already happened before industrialisation pushed large amounts of CO_2 into the atmosphere.

We live on a dynamic planet: things are always changing. Green places dry out, dry places get greener (as is happening to much of the Sahara Desert, but I'll bet you didn't see *that* in the mainstream media). The planet warms or cools. Sea levels rise or fall (and they are rising now no faster than they have for the past 8,000 years—and much slower, in fact, than for most of that time). On a planet as large and diverse as Earth, something remarkable will be happening somewhere every day—probably every minute; it is easy as pie to seek out odd events and blame them on whatever you want to demonise. But the dangers of putting the blame where it does *not* belong are, we shall see, deadly.

4
Is Carbon Dioxide a Major Cause of Warming?

Unfortunately, many experts who should know better seem to be arguing something like this: "Things are changing, we can't find any reason why they should be, so it must be caused by carbon dioxide emitted by human beings." For example the IPCC said this:

"During the past 50 years, the sum of solar and volcanic forcings would likely have produced cooling. Observed patterns of warming and their changes are simulated only by models that include anthropogenic forcings."[80]

This is not valid scientific reasoning, but it is very understandable that it seems persuasive. For example, if we leave a puppy alone in a room and come back later to find the rug chewed up, we might argue that the puppy must have done it because no one or nothing else could have. In many everyday situations that is a reliable way to think. But this is because we have a good understanding of the situation. We know that the other inhabitants of the room—the television, the clock, the sofa, for example—do not and can not chew rugs; we know that the windows and doors were locked and were not broken when we returned. Our understanding of rooms and clocks and puppies is good enough to encompass all the circumstances of the case, and we can reliably reason that the puppy must have chewed the rug.

This is not the case with climate science. When something as complex as the climate of a planet is involved, all kinds of mechanisms and effects could be operating that one simply hasn't thought of checking out or, even if we have thought of it, we do not have the means to do so. This is known to be the case in climate science. The models do not include all the factors operating in the physics of the

---

80   http://www.ipcc.ch/publications_and_data/ar4/syr/en/spms2.html

real atmosphere, and they simplify most of the ones they do include.

Perhaps the IPCC and those who say likewise would feel I have misrepresented them; "We have considered many other explanations," they might say, "and none of them are sufficient to explain the temperature rise." On this score, one of the world's greatest physicists, Freeman Dyson, says:

"I have studied the climate models and I know what they can do. The models solve the equations of fluid dynamics, and they do a very good job of describing the fluid motions of the atmosphere and the oceans. They do a very poor job of describing the clouds, the dust, the chemistry, and the biology of fields and farms and forests. They do not begin to describe the real world that we live in. The real world is muddy and messy and full of things that we do not yet understand. It is much easier for a scientist to sit in an air-conditioned building and run computer models, than to put on winter clothes and measure what is really happening outside in the swamps and the clouds. That is why the climate model experts end up believing their own models. ...

"When I listen to the public debates about climate change, I am impressed by the enormous gaps in our knowledge, the sparseness of our observations and the superficiality of our theories. Many of the basic processes of planetary ecology are poorly understood. They must be better understood before we can reach an accurate diagnosis of the present condition of our planet. When we are trying to take care of a planet, just as when we are taking care of a human patient, diseases must be diagnosed before they can be cured. We need to observe and measure what is going on in the biosphere, rather than relying on computer models. ...

"The biosphere is the most complicated of all the things we humans have to deal with. The science of planetary ecology is still young and undeveloped. It is not surprising that honest and well-informed experts can disagree about facts."[81]

Given this, perhaps it is somewhat arrogant to assume one knows and has taken account of all the relevant influences. The IPCC's mistake possibly needs a name; in fact it does: *argumentum ad ignorantiam*, or the appeal to ignorance. Perhaps it should be called "the fallacy of proof by lack of imagination."

---

81 Freeman Dyson. *Many Colored Glass: Reflections on the Place of Life in the Universe (Page Barbour Lectures).* University of Virginia Press, 2007.

The Case Against CO_2

We need to be clear just what sort of evidence is given to show that CO_2 causes significant warming. It is probably true that most people uninitiated into the details of the debate imagine that some kind of detailed understanding has been obtained showing exactly how and why carbon dioxide is now doing something dastardly and essentially different from how it has acted throughout the planet's history. As Freeman Dyson explains in the quote above, this is not the case.

The entirety of the case for the villainy of CO_2 is unverified computer models. These models do *not* simulate the actual physics of the climate in the sense that, say, a flight simulator mimics the real physics of a real airplane. The real planet is too complicated, and probably the only good simulation is the real thing. In fact, all the models make assumptions before the first line of computer code is ever written. Then they use those assumptions in building the model: the code is written so that it works according to the assumptions. Even if we knew all the relevant factors, the models could not be written to work according to the real cause and effect because the computer time needed would be practically unbounded, due to the complexity of the problem. J. Scott Armstrong, a professor at The Wharton School, University of Pennsylvania and a leading figure in the discipline of professional forecasting, has identified 140 principles of forecasting which must be followed to get reliable forecasts. Armstrong and Kesten C. Green of Monash University conducted a "forecasting audit" of the IPCC Fourth Assessment Report and found it violated no fewer than 72 of these principles.[82]

What this means is that, for all the fine-sounding talk about science and 'climate scientists', the most generous thing we can say about such models is that, *if* the assumptions of the model designers are correct, then *maybe* the output of the model *might* bear some resemblance to what *might* actually happen in the real world. Even if the assumptions are right, the model might still go wrong, but if the

---

[82] Craig Idso and S. Fred Singer, *Climate Change Reconsidered*. 2009 Report of the Nongovernmental International Panel on Climate Change (NIPCC).

assumptions are wrong, the model is completely worthless.

Ultimately, then, the case against CO_2 amounts to this: models that assume at the outset that carbon dioxide causes warming indeed go on to predict warming; then, because we "know" warming is "happening" and we can't think of any other cause for it, we decide that the cause must be carbon dioxide.

This reasoning fails on every step:

- As we have seen, the warming that has happened, although not completely absent, has been much exaggerated;
- other possible causes of warming have been proposed and largely ignored by the political IPCC—we shall look at these shortly;
- even if we could not think of any other reason, that doesn't mean there isn't one; and
- the models have not, in fact, made accurate predictions.

Failed Predictions—Warming Theory Refuted

For most of the twenty-first century, the warming of the four previous decades seems to have stopped or reversed. Oceans are losing heat, sea levels are now static or nearly so. We mustn't forget that weather isn't climate, and a cold or hot season doesn't, on its own, tell us much; but nothing in the climate models explains this—excepting for those models that have been tinkered with after the fact!

The Atmospheric Hotspot

Another test of the models is the kinds of changes they predict that the atmosphere should be undergoing. This is a critical test of the greatest importance, and if the models fail it, we can be virtually certain that the physical assumptions built in to these models are erroneous. It needs careful examination to see why, but the effort taken is well worth it, as it more or less clinches the case; and you don't have to be one of the high priesthood (a climate 'scientist') to know the truth for yourself.

When I started looking into the claims of dangerous warming due to carbon dioxide, I was at first baffled, buried in details of climate models, puzzled by energy balance diagrams, and so forth. Was there a "greenhouse" blanketing the Earth, slowly frazzling us to death? The truth could have been anything. If you've followed this path too, you'll know what I mean. But one thing, one single piece of the jigsaw, cut through all the fog and answered the question. I want to show you the thing that clinched the global warming question for me. I have postgraduate training in physics, which helped, but the basic point is understandable by anyone, and I want to explain what seems to me a key, conclusive fact in everyday terms.

To see why the key evidence (which I'll come to shortly) is so important, we need to have some 'physics intuition', so to speak. The following story about a forgetful child called Fred illustrates this.

Let's say it's a cold night and Fred climbs into bed (Figure 18).

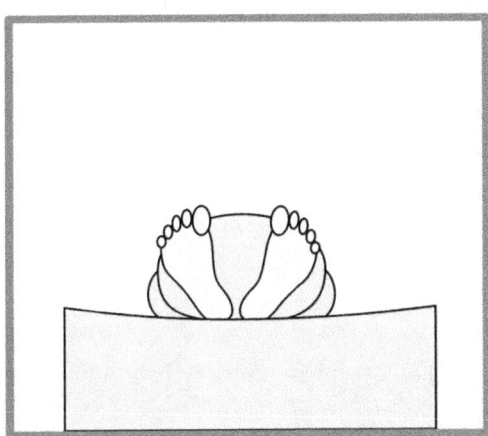

Figure 18: Fred in bed

Will Fred use a blanket to keep warm? If so, the air will heat up close to Fred because his body warms the air and the blanket prevents the warm air from moving away. On the other hand, as the night progresses, the air beyond the blanket will cool. In Figure 19 the "+" and "-" signs show air that becomes warmer or cooler respectively.

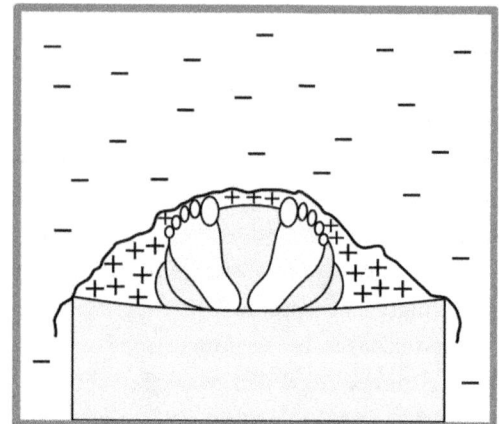

Figure 19: A blanket holds hot air close to Fred.

Now what if Fred (forgetful Fred) didn't use a blanket? The warm air escapes and tends to rise, since warm air is less dense than cold air (Figure 20).

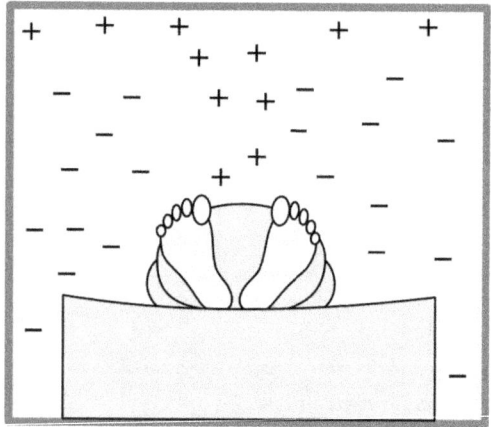

Figure 20: With no blanket, warm air escapes and Fred shivers

Poor Fred gets colder as the night wears on. But now we come to the point of the exercise: How do we know whether Fred used, or did not use, a blanket?

In real life, of course, we might simply take a look. But for the sake of our investigation, let's suppose that Fred is a very light sleeper, we dare not put on the light, so there's no way we can see if

Carbon is Life

there's a blanket. But we just happen to have an infra-red scanner that can tell us the temperature of the air at various spots throughout the room. Depending on whether Fred uses a blanket, the temperature change in the room follows one of the two characteristic patterns, as we saw above; so if we check where the air gets colder and where it gets warmer as the night wears on, we know for a fact whether or not Fred used a blanket, even without being able to see it. If Fred *did* use a blanket, our scanner should show results as in Figure 21; note that we can't see the blanket, but we can be sure that it is there.

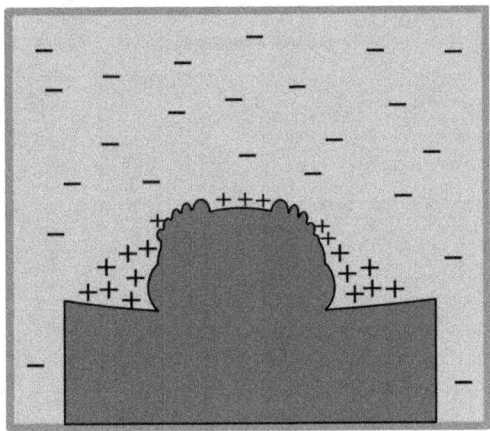

Figure 21: Warm air collects in a contained region, so there must be a blanket

On the other hand, if Fred does *not* use a blanket, we will see the temperature change in a pattern something like that in Figure 22.

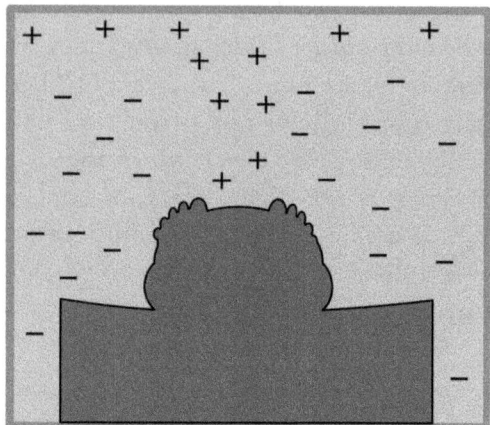

Figure 22: Warm air escapes upwards, so we can be sure there is no blanket

Once again, there is no doubt at all what is going on. In science, nothing is absolutely certain, but depending on which temperature pattern develops, we can be very, very sure indeed of the answer to the question: Did Fred use a blanket?

This illustration shows that in any situation governed by a physical principle, we see characteristic patterns developing. To take another example, fluid flow: here we see characteristic patterns such as smooth fluid flow, turbulence, vortices, and so on, and these develop when the system satisfies certain criteria. If we know the correct principles governing the case in hand, we can predict the nature of the patterns which we see.

Aircraft designers, for instance, can use computer modelling to calculate where airflow will be smooth, or where it will be turbulent, and what sort of turbulence patterns will be seen, and thereby predict the performance of an aircraft design. Experiments in wind tunnels are then used to fully test mockup models that look promising in the computer simulations. But if things don't happen as predicted, something is wrong: something has been calculated wrongly or some physical principle has been misinterpreted. The very last thing we would expect is that an aircraft designer, testing a model in a wind tunnel and seeing something completely at variance with the computer modelling, nevertheless shrugs his shoulders and suspects nothing is amiss.

But what about global warming? If the Earth is surrounded by a 'blanket' of anthropogenic (human-generated) greenhouse gas stoking up the temperature of the planet, then certain very characteristic patterns must emerge. The physics of a real blanket (as with Fred in the fable above) and a gaseous 'blanket' around the Earth differ, but that doesn't alter the fact that different heat dissipation (or retention) processes will result in different characteristic patterns of temperature change. Just as Fred will be surrounded by something roughly resembling one of two quite different patterns of air temperatures, so likewise will temperature changes around the Earth have a quite definite pattern, depending on which climate theory is right. Scientists whose pay check does not depend on agreeing with global warming alarmism will all agree with this simple statement. It is part of the basic skill of having a 'nose' for physics.

What, then, are our main competing climate theories? The IPCC's reports are based on results from a collection of climate computer models; they have *nothing* else. These are simply computer programs that, in essence, contain a computerised version of the assumptions and beliefs of the climate modellers as to how the climate of the planet works. Whether these assumptions are well-founded is another question, but the key point is that whatever these assumptions may be, when the climate model is run, it generates its 'predictions' by calculation of hypothetical futures for the behaviour of the atmosphere. These 'futures' necessarily contain predictions of the changes of atmospheric temperatures at various heights above the planet and the various latitudes all the way from south pole to north pole.

The indisputable fact about these atmospheric temperature predictions is that if the pattern doesn't happen, the model is wrong. Just as Fred won't warm up if he isn't surrounded by warm air, or an airplane won't fly if the correct pattern of airflow upon the wings isn't there, so likewise the effects on the Earth of global warming cannot happen if the cause of the warming —the warm air—isn't there.

So now we come to the graphs that clinch the matter. All global warming models predict some sort of developing 'hotspot' in the atmosphere above the tropics. Figure 23[83] shows the graph for one of the models, but they all look roughly similar.

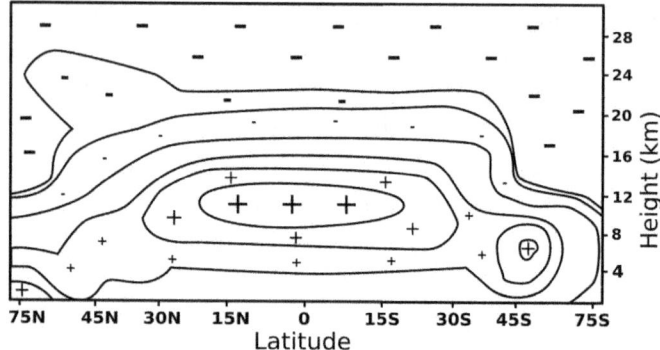

Figure 23: Atmospheric temperature changes as predicted by a climate model ('+' shows warming, '-' shows cooling; period 1958 – 1999). Note the tropical 'hotspot'.

This picture shows the atmospheric temperature trends as predicted by the computer model from 75 degrees north to 75 degrees south (the equator in the middle) and up to 30 km above the Earth. We can think of this air pattern as corresponding to the pattern in Fred's bedroom when Fred used a blanket. Just as we did with Fred in bed, we can compare reality with this picture. Is the heat in the real atmosphere doing what the model predicts? Figure 24[84] shows the measured temperature trend in the real world.

83 Adapted from the NIPCC Report page 107.
84 Adapted from the NIPCC Report page 106.

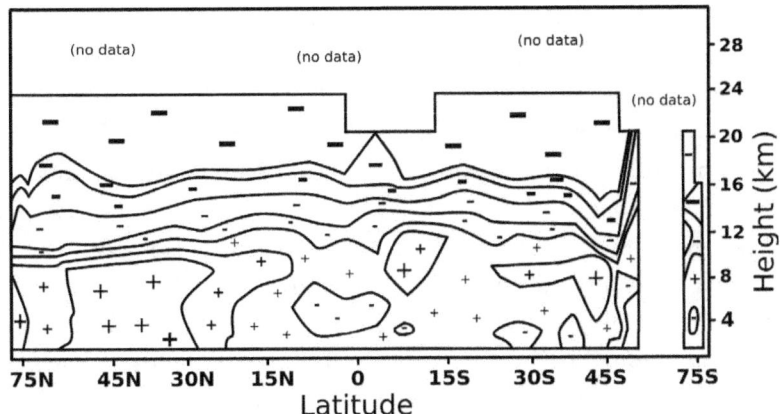

Figure 24: Measured data shows a completely different temperature trend, which has no hotspot.

What have we actually proved here? Well, *proved*, without any possibility of error, nothing, of course; no serious question about the real world ever has a complete and perfect proof as an answer, so don't be misled if someone says the world still *might* be heating due to CO_2 despite the absence of the warm spot that is supposed to do the warming. Of course anything *might* happen. Milkshakes *might* rain down out of the sky; but how likely is it? How likely is it that Fred has a blanket, but the air around him is getting colder just as if he had no blanket, and yet Fred is warming up despite that? The two questions have the same answer: not very.

Yet surprisingly, some proponents of global warming alarmism actually resort to this very strategy. "True," they say, "the hot spot isn't developing. But that is because the heat is being *stored up* elsewhere—it's 'in the pipeline'—and one day it will burst forth with even greater severity and vengeance."

What can we make of that claim? Well, thinking back to Fred again, it amounts to this: We use our temperature probe in Fred's darkened bedroom and we see a pattern like that in Figure 22 above, corresponding to no blanket: Fred should be freezing! But actually, the heat has all gone into Fred's body, despite the complete absence of the hot air which is the mechanism for making it do so. In other words, Fred got warmer by disobeying the second law of thermodynamics—in other words, by magic. Likewise, if someone says heat is being secretly stored somewhere by global warming, despite the absence of the very mechanism that does the warming, they are saying global warming is happening by magic. That is the harsh truth of the matter.

Like many other physical systems, there are things about the climate that can be understood by any intelligent person willing to learn some simple physics. The 'hotspot' is one of them. All scams depend on persuading people to trust the hoaxsters rather than their own good sense; the global warming scam is no different.

It is important to remember that there is no direct observational evidence that carbon dioxide causes dangerous warming—in fact there is counter-evidence. It all hinges on computer models. Here are some other things the models are getting wrong:

- The models failed to predict the falling off of sea level rise (which never accelerated beyond its background rate anyway);
- They failed to predict the levelling off of temperature since around 2000;
- They failed to predict the recent incredibly snowy winters across the northern hemisphere; in fact the key warming advocates claimed that snow would become a rarity;
- They predicted vastly increased hurricane activity, yet hurricane energy is close to its lowest in 30 years.

In sum, the models are batting zero: their predictions are no better than random guesses.

It is hard to believe that things are so bad in a field that claims to be a science. In discussions with friends, I find they are incredulous that so much has gone so very wrong. Surely, they ask, there must be *some* reason why the models are based on the assumptions they are? And what are those assumptions? Why are they so confident despite not having any predictive success? To get some insight into this, we need to look at the physics.

The Physics of Greenhouse Gasses

This is important to fully appreciate the significance of the missing hotspot. If you don't have a science background, it might not be apparent that the missing hotspot is pretty much proof positive that the global warming theory is completely and absolutely wrong. To get a 'feel' for this, to make it 'come alive', we need to look at the concrete physical reason for the hotspot.

The behaviour of the climate models (and most others) hinges critically on the correct handling of a physical process known as feedback. What happens if a certain amount of extra carbon dioxide is placed into the atmosphere? CO_2 is a greenhouse gas, so a certain amount of warming happens. Exactly how much is a matter of debate, but there is near-general agreement that without any of these gasses, the world would be at least twenty, possibly thirty Celsius colder and would be a frozen iceball. Based on the IPCC's own figures, the warming due to a doubling of CO_2 would be about 1.2 degrees Celcius.[85] Others give figures such as 0.9 C, 1 C, and so on. But 1 C, or 1.2 C, wouldn't be harmful in the slightest; it is much more likely on the whole to be beneficial.

But then there's feedback. The relevant fact here is that CO_2 is not the major greenhouse gas, not even remotely so. The gas with by far the greatest greenhouse effect is water vapour. Unfortunately for the politicians, traders, and so on supporting the greenhouse alarmism, you cannot tax or trade in water vapour. So instead the story is told like this: CO_2 causes some warming so, due to the resulting higher temperature, more water evaporates from the oceans, raising the greenhouse content of the atmosphere, which in turn causes more warming than the CO_2 alone would have caused. The feedback warming due to the extra water is claimed to be much more than the original warming due to the carbon dioxide; exactly how much depends on the model.

There are some odd things about this theory. Firstly, we know

---

85   http://wattsupwiththat.com/2012/04/23/why-there-cannot-be-a-global-warming-
       consensus

that in the geological past, the planet has been frozen solid, all over, with atmospheric CO_2 many times greater than it is today. If CO_2 caused critical positive feedback, then surely, we might ask, would not the planet have suffered irreversible heating long before now? Human-produced CO_2 is only 3% of all CO_2 emissions. Volcanoes, comets, nearby supernovae, all these things have affected Earth some time during its 4.3 billion year history. And we are asked to believe that none of them, at any time in this vast history, ever made a small change such as humans are now making, thereby tipping the planet into a runaway greenhouse? Is the Earth so delicately balanced, like a pencil standing on its point, and yet it has survived 4.3 billion years without *anything* tipping it over? This doesn't pass the 'smell' test!

The next problem with the greenhouse theory is how it is supposed to work. Our major source of energy is the Sun, and because the Sun is much hotter than the Earth, most of the energy in sunlight is in the visible and ultraviolet parts of the spectrum. These wavelengths pass unobstructed through the atmosphere and hit the Earth, thereby heating it. The Earth then radiates that energy back outwards, but, due to the Earth's lower temperature, it radiates mostly in the less-energetic infra-red part of the spectrum. But somehow or other, all the energy arriving from the Sun has to be radiated back into outer space so as to balance the amount of heat arriving with the amount leaving, or else the planet would get hotter and hotter.

Water and carbon dioxide both absorb energy in parts of the infra-red spectrum. Water absorbs at many frequencies, but CO_2 mainly absorbs in just one band. This is why water is a much more potent greenhouse gas. But here we run into a big problem for the theory: just how much of the energy that CO_2 *might* absorb, *is* it absorbing?

The answer is: all of it. At existing concentrations, CO_2 is already absorbing virtually all of the photons in its bandwidth. Such a photon emitted from the surface of the Earth will not get more than *ten metres* from the surface before being absorbed.[86] And if CO_2 con-

---

[86] http://www.john-daly.com/artifact.htm

centrations were, to pick a wild number, multiplied by five? (In fact there isn't enough fossil fuel in the world to do this.) Then the photons would be absorbed within two metres instead of ten metres. But two metres or ten metres, the photon is some tens of *kilo*metres short of escaping the atmosphere into space. That is, more CO_2 won't make any significant difference.

Radiating to space at these wavelengths cannot happen at or near ground level because the air above would just absorb the photon again. Neither two nor ten metres solves the problem of getting that energy tens of kilometres up into the sky in order that it can be radiated to space.

The answer is that the heat is *not* radiated upwards, it is convected, just as in a saucepan of water on a stove, the water circulates. The stove heats the bottom of the saucepan, the hot water rises, and cold water falls down so that it, too, can be heated. The same thing happens in the atmosphere, and anyone except a climate scientist locked in his lab with his faulty computer models can see it for themselves. Hot air at ground level rises, taking the heat with it, and cool air from above falls down to heat up in its turn.

I live in the country, and from time to time, my bird friends yell and squawk to call me outside when there is danger afoot; sometimes it is a snake or a goanna; and sometimes it is the amazing spectacle of huge wedge-tailed eagles circling on a 'thermal'—rising hot air—using the energy of the air column to lift the eagle to great heights.

But if you watch an eagle closely, you will see that they glide in a circle that is only a few hundred metres across. These thermals, one of the major heat transport mechanisms, can be comparatively small things. The climate models, on the other hand, cannot hope to model such small features, typically dividing the planet into 'cells' that are at a minimum some kilometres across. If the CO_2 content were doubled, then the photons would be absorbed within five metres of the surface rather than ten metres; they would still heat the air, and still power the thermals, and we could all go on admiring the magnificent sight of a proud eagle ascending into the heavens. And the heat would still escape. But the climate models, unable to work on such small features, have to be given 'parameters' (a fancy name for guesses) to handle all this wonderful interaction of life and nature. No one has a magical intuition that tells them the right values for the parameters, so guesses invented by climate 'scientists' are no better than yours or mine. Indeed, insofar as you and I know our guesses are worthless, we are one step ahead of the climate 'scientists'.

We can now 'intuit' the importance of the missing hotspot. The global warming theory has it that as the extra CO_2 causes a little bit of extra warming, that little extra warmth causes more evaporation of water from the oceans. That vast rising volume of water vapour (water being a much more potent greenhouse gas than CO_2) then causes a *lot* more warming. *And that's the hotspot!* No hotspot means no extra warming, which means the theory is wrong. It's so simple anyone but a climate scientist could understand it. Again, anyone who thinks the warming is happening anyway is saying global warming happens by magic.

But the problems for the theory don't stop there. The next problem is clouds. The climate models do not understand, nor correctly account for cloud formation—all that extra water is cloud food when it rises and cools. Just 1% more cloud can overwhelm all the effects of extra CO_2. Have they, or even, can they, correctly model cloud formation? The modellers themselves admit they can not; the size of the uncertainty is larger than the entire calculated value of the feedback[87]—which tells you once again, if any more proof were needed, that the models are useless.

The Models Falsified—The Theory Refuted—Again

We have seen that the models cannot account for many critical heat dispersal mechanisms used by the real planet to cool itself; but the nail in the coffin of the climate models is the fact that observation

---

[87] Kurt Lambeck, professor at the Australian National University's research school of earth sciences, has admitted this. www.theaustralian.com.au/ national-affairs/ humans-affect-climate-change/story-fn59niix-1225906508258. August 18, 2010

has now falsified them.

The observational test arises from this key fact: if warming due to carbon dioxide causes positive feedback from water vapour, then as the temperature rises, the energy radiated into space must fall compared with what would be radiated without the positive feedback. The experiment has now been done, and the feedback is negative, not positive. Earth radiates much more energy as it heats than it would if it were an inert body: it has a thermostat.

Lindzen and Choi compared the predicted outgoing radiation with that actually measured:

> "Climate feedbacks are estimated from fluctuations in the outgoing radiation budget from the latest version of Earth Radiation Budget Experiment (ERBE) nonscanner data. It appears, for the entire tropics, the observed outgoing radiation fluxes increase with the increase in sea surface temperatures (SSTs). The observed behavior of radiation fluxes implies negative feedback processes associated with relatively low climate sensitivity. This is the opposite of the behavior of 11 atmospheric models forced by the same SSTs. Therefore, the models display much higher climate sensitivity than is inferred from ERBE..."[88]

This is dramatically demonstrated in Figure 25[89], modelled on one from Lindzen & Choi's paper. In the models, extra temperature *restricts* heat flow and *increases* the feedback; in the real world, extra temperature *increases* heat flow and *decreases* the feedback. Remember, the increased feedback in the models was not output *by* the model after simulating the real world physics, it was an input *to* the model based on what the modellers guessed would happen—and as the figure shows, they could have done better by just rolling dice. In a nutshell, this shows us that the entire global warming scare is built on a mistaken guess.

---

88 Richard S. Lindzen and Yong-Sang Choi, *On the determination of climate feedbacks from ERBE data*. Geophysical Research Letters, VOL. 36, L16705, doi:10.1029/2009GL039628, 2009.

89 This diagram is modelled on a presentation of Lindzen and Choi's results as given in a public talk by Viscount Christopher Monckton in Brisbane 29 Jan 2010.

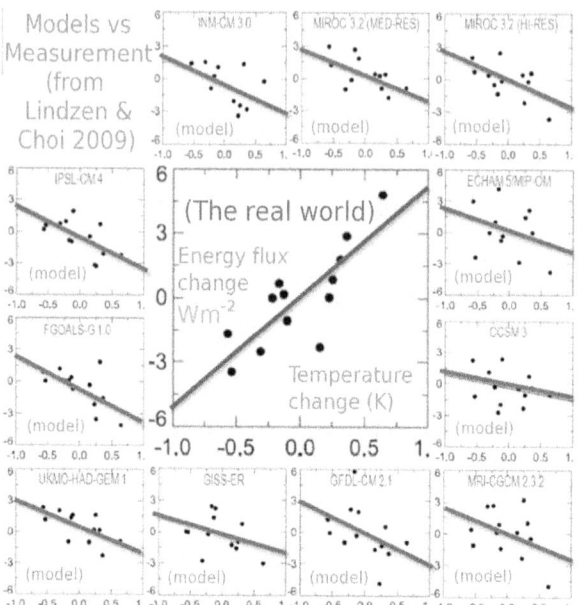

Figure 25: Every model gets the flux the wrong way round

And the GW Theory Refuted—Yet Again

The catastrophic human-caused global warming theory—aka 'climate change' (though the climate has *always* changed!) aka 'global warming' (though the climate has either warmed or cooled since the planet was created) must surely be the most thoroughly refuted theory in the history of science. And now Ken Gregory has publicised some NASA data that puts yet another stake through the heart of this monster.

Recall that, in brief, the theory goes like this: everyone agrees that CO_2 causes some warming—about 1C for each doubling of the CO_2 concentration. That would be quite harmless, but the theory goes on to say that this small warming by the CO_2 causes water vapour to evaporate from the oceans. The warming due to this water vapour multiplies the effects of the CO_2 alone, resulting in a dangerous amount of warming. We have seen that this theory predicts a hotspot in the atmosphere, which simply isn't there. That fact, alone, disproves the theory. But it turns out that there is actual data that directly shows that the water vapour concentration itself has been decreasing—the precise opposite of the theory's prediction. The NASA water vapour project (NVAP) has directly measured changes in water vapour in the upper atmosphere, and it is going down, not up.[90] Gregory writes:

"Climate models predict upper atmosphere moistening which triples the greenhouse effect from man-made carbon dioxide emissions. The new satellite data from the NASA water vapor project shows declining upper atmosphere water vapor during the period 1988 to 2001. It is the best available data for water vapor because it has global coverage. Calculations by a line-by-line radiative code show that upper atmosphere water vapor changes at 500 mb to 300 mb have 29 times greater effect on OLR and temperatures than the same change near the surface. The cooling effect of the water vapor changes on OLR is 16 times the warming effect of CO_2 during the 1990 to 2001 period."

The theory says water vapour goes up and this makes temperature go up by reducing the escape of heat to space. We have now seen that there is direct physical evidence disproving every step in this logic: The water vapour goes down, not up, the heat escape to space goes up, not down, and the 'hotspot' is cool, not hot! As a scientific theory, 'climate change' is dead. If it lives on, it is because it is kept alive by considerations that are alien to the spirit and methods of genuine science.

Problems with the Models Keep on Coming

Once one understands that the models do not *discover* a positive feedback from CO_2, but instead have it *built in* as a starting assumption, then much of the awe for their 'science' dissipates. The models don't generate their results by applying the basic physics of all of the processes happening in the real world; they address only a subset of

---

[90] Water Vapor Decline Cools the Earth: NASA Satellite Data. Ken Gregory. March 4, 2013. www.friendsofscience.org/assets/documents/NVAP_March2013.pdf. Reposted at http://wattsupwiththat.com/2013/03/06/nasa-satellite-data-shows-a-decline-in-water-vapor.

these processes in making a model, and augment it at the outset with question-begging assumptions to deal with the factors the model can't handle. So how can any 'findings' come out of those models?

About the IPCC report's models, the NIPCC Report, a response by critics of the IPCC Report, writes:

> [Two experts in forecasting commented:] The forecasts in the [IPCC] Report were not the outcome of scientific procedures. In effect, they were the opinions of scientists transformed by mathematics and obscured by complex writing. Research on forecasting has shown that experts' predictions are not useful in situations involving uncertainty and complexity. We have been unable to identify any scientific forecasts of global warming. Claims that the Earth will get warmer have no more credence than saying that it will get colder. ...
>
> Prof. Freeman Dyson, one of the world's most eminent physicists, said the models used to justify global warming alarmism are "full of fudge factors" ...
>
> Dr. Antonino Zichichi, emeritus professor of physics at the University of Bologna, former president of the European Physical Society, and one of the world's foremost physicists, said global warming models are "incoherent and invalid."[91]

This highlights a central problem with the use of these models: failure to follow the scientific method. The IPCC is very careful to disclaim that they are making "predictions" with their "scenarios" in their Reports. That just isn't good enough. Science is all about falsifiability: if you disclaim making a prediction, if you refuse to accept that there are any tests that might disprove the soundness of your work, then whatever else it might be, it isn't science. If you disown criticism when your "scenario" doesn't come to pass, equally you cannot accept any kudos if your "scenario" happens to come true (not that they have so far). And you certainly cannot talk about the assumptions of the model as being the "only explanation".

Yet another problem with these models is that their reputed accuracy in making 'hindcasts' is undeserved. A hindcast is using the model, starting some time in the past, to "predict" the temperatures from that time onwards. Because the model run started in the past, the "predictions" are for times that, in the real world, have already happened and so we can compare the model results with reality. An

---

91   The NIPCC Report, pages 10, 11.

excellent piece of scientific investigative journalism will be found in a blog post by Willis Eschenbach[92]. By replicating a GISSE model hindcast with a model of his own, he was able to show that the modellers had tuned parts of the model's input to allow the model to produce the required temperature plot, and that the tuned forcings[93] were not the same, in some cases, as the known historical facts. For example, the model assumed that the world's black carbon emissions (smoke, forest fires, etc.) had continually increased throughout the twentieth century, when in fact they dropped off significantly after about 1950.

In other words, the model's "correct" hindcasted temperature plot was made using fantasy figures for the behaviour of some of the forcings. That is, it was applied to 'some other planet'! The model's impressive temperature hindcast might make good publicity, but it is scientifically worthless. When you see astounding claims being made in the press about new 'investigations' into global warming, you can be reasonably sure the reporter will not have done his or her homework in the impressive manner of Eschenbach's. Time and again in the global warming business, startling press reports turn out to be nothing more than flim-flam.

Eschenbach relates this in his own humorous style to the "no other explanation" argument as follows:

"Here is the sting in the tale. They have designed the perfect forcings, and adjusted the model parameters carefully, to match the historical observations. Having done so, the modelers then claim that the fact that their model no longer matches historical observations when you take out some of their forcings means that "natural forcing alone cannot explain" recent warming ... what, what?

"You mean that if you tune a model with certain inputs, then remove one or more of the inputs used in the tuning, your results are not as good as with all of the inputs included? I'm shocked, I tell you. Who would have guessed?"

---

92 http://wattsupwiththat.com/2010/12/19/model-charged-with-excessive-use-of-forcing

93 Forcings are things that are input to the model and are beyond its control, and include things like land use changes, volcanic eruptions, changes in solar output, and so on.

Hard Data Reverses the Temperature-CO_2 Link

We have just seen that not only do the models fail to predict the climate correctly, they assume at the start the 'findings' that they output at the finish. But that's not all. After all, why do we keep on getting blitzed with news stories about hottest year, hottest summer, and so on? Because we assume (unless we know better) that CO_2 is the cause and temperature is the effect. After all, ice core data taken from Antarctica and also Greenland shows conclusively that for the past several million years, carbon dioxide concentrations and temperature have been in lock-step. Surely CO_2 must be the villain? The only problem is, the temperature changes first and the carbon dioxide changes some six to eight *hundred* years later. In other words, temperature change causes more outgassing of CO_2; CO_2 does not cause temperature change.

On the really long term, the approximately 645 million years since the explosion of multicellular life in the Cambrian, and beyond, right back to the formation of the Earth 4.5 billion years ago, there is no correlation between atmospheric CO_2 concentration and temperature.

Statistics Rips the Climate Models to Shreds

A startling new paper appeared in 2012. Beenstock, Reingewertz, and Paldor have published an advanced statistical analysis to test the global warming (AGW) theory over the time period 1880-2007[94]. They find that greenhouse gas forcing, aerosols, solar irradiance and global temperature are *not* related according to a powerful statistical test called polynomial cointegration. This means that the relationship climate researchers think they have discovered between these variables is an illusion, an accident of the way they have analysed the

---

[94] Beenstock, M., Reingewertz, Y., and Paldor, N.: Polynomial cointegration tests of anthropogenic impact on global warming, Earth Syst. Dynam., 3, 173-188, doi:10.5194/esd-3-173-2012, 2012. Online at http://www.earth-syst-dynam.net/3/173/2012/esd-3-173-2012.html

data. They write:

"The fact that since the mid 19th century Earth's temperature is unrelated to anthropogenic forcings does not contravene the laws of thermodynamics, greenhouse theory, or any other physical theory. Given the complexity of Earth's climate, and our incomplete understanding of it, it is difficult to attribute to carbon emissions and other anthropogenic phenomena the main cause for global warming in the 20th century. This is not an argument about physics, but an argument about data interpretation. Do climate developments during the relatively recent past justify the interpretation that global warming was induced by anthropogenics during this period? ... [O]ur results challenge the data interpretation that since 1880 global warming was caused by anthropogenic phenomena."

They explain: "The implication of our results is that the permanent effect is not statistically significant" and, in the typically cautious academic manner, add that "scientists who make strong interpretations about the anthropogenic causes of recent global warming should be cautious. Our polynomial cointegration tests challenge their interpretation of the data."

They do, however, find that there could be a temporary warming effect due to human activity. They write:

"If the effect is temporary rather than permanent, a doubling, say, of carbon emissions would have no long-run effect on Earth's temperature, but it would increase it temporarily for some decades. Indeed, the increase in temperature during 1975–1995 and its subsequent stability are in our view related in this way to the acceleration in carbon emissions during the second half of the 20th century... . The policy implications of this result are major since an effect which is temporary is less serious than one that is permanent."

What does this mean in plain words? My take on it hinges on the difference between what the climate "scientists" have done and what Beenstock's group have done. The climate models are computer programs that have various assumptions built into them at the outset. Next, they are numerical calculations, meaning they have errors and will inevitably "drift" (meaning, get increasingly inaccurate) the further into the future they are computed. This is a necessary feature of all such numerical simulations. This can go wrong at many places. The inbuilt assumptions could be wrong (and we have just seen that they are wrong for sure). Next, the numeric simulation could be too "coarse": not enough grid points to capture reality—and we know this is true about them too. Also they might not model all the processes that are actually occurring on the real planet—and we know this for a fact as well. The climate models are looking pretty threadbare by now.

And after all this, the modellers cannot get a "fit" to the world temperature history without assuming an AGW effect. We would be perfectly entitled to reject the entire AGW claim just on the weakness of this 'evidence'. But here's something we can bank on: if the modellers are right, and an AGW affect exists, it must betray its presence statistically, because if it doesn't, it must be random, and if it's random, it isn't a connection between human activity and global warming.

In a nutshell, Beenstock's team has used one of the most powerful statistical tests known, and found that, *at the most*, there can be a temporary warming effect, but almost certainly not a permanent one. As they say, a temporary effect is much less serious than a permanent one. So, even if the scariest claims about the plant food, carbon dioxide, are correct, why would one think that introducing crippling "carbon reduction" legislation (killer power and fuel costs and caps and taxes on energy usage, meaning on virtually everything that makes for a liveable modern society) will be good value for money? On even the most optimistic estimates, drastic actions like Australia's carbon tax will only lower temperatures by a few thousandths of a degree by 2050. Impoverishing people and denying the planet the life-giving plant food that is CO_2 would be bad enough if that were a permanent few thousandths of a degree; but if it is only a temporary saving for a decade or two, continuing with such policies must surely be lunacy. Wouldn't it be better to spend far less money far more effectively dealing with any problems as they occur (especially since the scare stories might *never* occur)?

5
Lots of Things that Have Been Overlooked

We have seen in the previous Chapter that the catastrophic human-caused global warming theory is wrong: it has been disproved in multiple ways. As a scientific theory, it belongs in the history books alongside phlogiston, the mysterious 'fire substance' that was thought to cause fire before modern chemistry explained the real reason. Indeed, it is so bad it is more comparable to spontaneous generation, the discredited idea that mice and other living beings could grow spontaneously out of vapours or dirty rags. Yes, it is that bad.

As we saw in Chapter 1, carbon dioxide helps plants grow faster and the world has greened 6% in two decades as CO_2 concentrations have risen. That is food for a few hundred million people, mostly the world's poorest, who might have something to say about the perverse notion that we should base our policies on such bad science. *"Carbon reduction" is food reduction.* It starves the poor and it harms wildlife. The idea that we should do it anyway, "just in case" (a.k.a. the "precautionary principle") is trading a certainty for a near impossibility. We will look at that in depth in Chapter 8. Right now, we need to look at a whole slew of other results and discoveries that further reduce the credibility of the global warming theory.

Is the CO_2 Concentration Even Increasing?

Human-caused (anthropogenic) CO_2 emissions are a small part of a much larger process. Humans emit some 3% of the CO_2 released into the air each year, the rest coming from natural causes that have been going on since time immemorial. Whilst it is probably true that humans are contributing to the CO_2 rise, it might not be as dramatic an effect as we imagine.[95]

There doesn't seem to be much doubt that the atmospheric CO_2 concentration is indeed going up. Don't think, though, that we have any hard and fast facts about it. Unfortunately we can't trust anything we hear on climate 'science', not even the claimed historical temperatures[96], and not the history of atmospheric carbon dioxide concentrations.

Year	CO₂ ppm	Uncertainty (ppm)
1826	360	70
1840	340	120
1850	330	60
1880	310	10
1920	310	10
1940	380	5
1960	320	5

Table 3: Atmospheric CO_2 Background 1826-1960 shows higher values than the commonly accepted figure 280 ppm.

Modern measurements show CO_2 concentrations following a smoothly increasing curve (with seasonal wiggles) ramping up from the supposed pre-industrial figure of about 280ppm. But Ernst-Georg Beck has investigated over 200,000 proper scientific measurements of CO_2 concentration made by scientists since 1800, and he finds an entirely different story. Table 3 shows representative values taken from Beck's graphic summary of his research findings.[97] One might be

---

95 Paleoclimatology Professor Robert M. Carter, in his *Climate: The Counter Consensus* provides a summary of research based on the percentages of the various carbon isotopes in the atmosphere, indicating that the human contribution is less than we might intuitively expect. See pages 78-82.

96 Time after time, old records from the past are revised down, making the present appear hotter by comparison. For one shocking example, see: http://wattsupwiththat.com/2012/06/06/noaas-national-climatic-data-center-caught-cooling-the-past-modern-processed-records-dont-match-paper-records

97 Graphic at http://www.biomind.de/realCO2 based on Ernst-Georg Beck's paper:

tempted to dismiss such evidence simply because it is so hard to believe that so much is so wrong with the climate science establishment. But unfortunately they have a track record of distortions, untruths, lies, and intimidation and bullying (of journal editors, for example, and genuine enquirers asking for the basic stuff of science: raw data). Regardless of whether one finds Beck's work believable, the climate science establishment has blotted its copybook far too often to be automatically believed. Uncertainty pervades every aspect of the case against carbon dioxide, even to the point of casting doubt on whether it has increased at all since pre-industrial times.

And in Fact There Are Other Explanations

We have seen that "We can't find any other explanation" is bad thinking and isn't science; but it isn't true in any case. There *are* proposals for other causes apart from CO_2 to account for what warming (might) have happened.

Other Human Activities

It could be that warming has been caused by humans, but not by carbon dioxide. The basic laws of thermodynamics necessitate that human energy use generates heat. This heat would be emitted even if we solved the problem of fusion energy and converted the entire economy to a non-carbon base. It may be that if we want an advanced civilisation we have to bargain with having a slightly hotter planet, and nothing we do about carbon dioxide will change a thing.

Even wind energy might not be immune from this effect: power from wind will still be converted into heat when it is used, and the effects of draining energy from the wind might well reduce the planet's ability to transport heat efficiently upwards away from the surface.

Human beings do a lot more than put carbon dioxide into the at-

---

180 Years of atmospheric CO_2 gas analysis by chemical methods, Ernst-Georg Beck; Energy & environment Volume 18 No. 2 2007

mosphere. In fact, pretty much every other change we make is of more concern than CO_2, for the simple reason that emitting CO_2 is 100% beneficial to the planet, as we have seen. Here are a few things we really should be worrying about:

- deforestation: this directly drives heating by reducing surface moisture; also forests and other vegetation inserts oxygen into the atmosphere; forests are places for wildlife, which deserve their living spaces just as much as humans do.

- draining swamps: again, just because swamps might be really nasty places for humans doesn't mean they don't help other creatures; in many cases we simply don't know how much damage will follow from making wholesale changes like eliminating entire ecosystems; reported benefits of swamps include flood control, recharging of aquifers, filtering water, removing pollutants, and removing nitrate (which is dangerous to human babies); and needless to say, hundreds of species rely on swamps and wetlands.

- expanding cities and human occupation: cities are hotter than the countryside; cities and roads break up natural corridors for animals to travel and migrate; modern city planning tries to take this kind of thing into account, and hopefully we'll do even better at it in future; but the urban areas we have already created have done untold harm to the natural ecosystems previously in place; we need to better manage our usage of space for human purposes, not only for the effects on temperature but also for the many other natural mechanisms we disturb in the process.

- emissions of *genuine* pollutants: almost every other gas human industry emits is more of a problem than CO_2. Sulphur dioxide, nitrous oxides, ozone, and carbon *mon*oxide are major and dangerous pollutants, and some of them change the planet's temperature a little, either up or down depending on the gas. Unlike carbon dioxide, these chemicals really do cause serious harm to humans, animals, and the environment generally. Historical experience shows that we have become better at reducing these pollutants, and that political freedom and increasing wealth helps a society to do this; but wealth and freedom are the exact opposites of the cruel totalitarian proposals being put forward to fight the imaginary pollutant, carbon dioxide.

We have, therefore, a collection of mechanisms, some well understood, some less so or poorly understood (and perhaps some not yet discovered?) with which we might explain Earth's temperature changes. But for sure, we cannot blame whatever we don't understand on any one factor, be it carbon dioxide or not.

Cosmic Rays and Other Natural Causes

On a long time scale, the Earth's orbit changes slightly and this changes the received solar insolation (heat energy from the Sun). This is widely believed to be the main cause of the cyclic ice ages discussed earlier, but would not be responsible for the short-term climate changes causing worry at present.

More relevant are the Pacific Decadal Oscillation and El Niño/Southern Oscillation. These effects (the former on a time scale of decades, the latter on times around a year to eighteen months) result in varying patterns of heating and cooling (and winds and rainfall, etc.) over parts of the globe. They have to be taken into account in explaining temperature changes. But the most important effect is probably something quite different: the effects of cosmic rays upon the Earth's cloud cover.

Our sun is a variable star. It follows a sunspot cycle of roughly eleven years (but can be much longer, to around fifteen years). Over the course of each cycle, the Sun's sunspot activity increases from zero, causing much magnetic activity, and finally decreases back to zero. Then the magnetic polarity switches around and the next cycle repeats the pattern but with the opposite polarity, and on it goes. But the cycles are not all the same strength, and during the Little Ice Age, there was a time called the Maunder minimum when sunspots could not be observed at all. Telescopes had been invented by then, but the full range of modern instrumentation had not, so we don't know exactly how much solar activity there really was, but it was certainly very low and corresponded to extremely cold times on Earth.

In the opposite of that pattern, the last several cycles have been extremely active, and have been dubbed a 'grand maximum'. It is believed that for the fifty or so years until the last solar minimum (2008), the Sun had been more active than at any other time for the previous 8,000 years.[98] The reason solar activity is often discounted is that the Sun's energy output changes only a little during the solar cycles. How can a very small change make a big difference in climate?

A recent theory has it that the effect is indirect. Greater magnetic activity on the Sun deflects cosmic rays from hitting Earth. When the solar cycles are strong, very few cosmic rays penetrate to low altitudes. Conversely, weak solar cycles allow many cosmic rays to penetrate. The creator of this theory is Henrik Svensmark, a professor at the Technical University of Denmark. He writes:

> "When the Sun is active, its magnetic field is better at shielding us against the cosmic rays coming from outer space, before they reach our planet. By regulating the Earth's cloud cover, the Sun can turn the temperature up and down. High solar activity means fewer clouds and a warmer world. Low solar activity and poorer shielding against cosmic rays result in increased cloud cover and hence a cooling. As the Sun's magnetism doubled in strength during the 20th century, this natural mechanism may be responsible for a large part of global warming seen then."[99]

Svensmark's theory was tested in 2011 at Europe's CERN accelerator, and the critical step in the process was dramatically confirmed.[100] Cosmic ray action generates a steady stream of molecular clusters, believed to be the essential seeds for cloud droplets. The work isn't over yet, but the key obstacle to the theory has been re-

---

[98] S. K. Solanki et al, *Unusual activity of the Sun during recent decades compared to the previous 11,000 years.* Nature **431**, 28 Oct 2004 pp 1084-1087.

[99] http://wattsupwiththat.com/2009/09/10/svensmark-global-warming-stopped-and-a-cooling-is-beginning-enjoy-global-warming-while-it-lasts

[100] Well-known research physicist Nigel Calder has a very readable writeup on his site at http://calderup.wordpress.com/2011/08/24/cern-experiment-confirms-cosmic-ray-action .

moved.

A very distasteful postscript to this experiment is the reaction of Rolf-Dieter Heuer, Director General of CERN. One would expect a scientist to be enthusiastic about such revolutionary research at his own institution, But instead, he said (translated from the German): "I have asked the colleagues to present the results clearly, but not to interpret them. That would go immediately into the highly political arena of the climate change debate. One has to make clear that cosmic radiation is only one of many parameters."[101]

What are we to make of this? Do you recall *any* head of *any* establishment organisation warning his staff not to draw conclusions when their work favours the reigning global warming theory? Of course not. Yet here, where the results could not be clearer, where they pretty much blow global warming theory out of the water, where the fact that temperature marches in lock-step with the sun's magnetic activity is now explained and a complete alternative to carbon dioxide as climate driver is plain for all to see, this manager tells his staff to say nothing about it! When we are about to wreck modern civilisation by closing down cheap energy, the fact that there is another explanation for climate changes is surely the most important single fact that should be publicised far and wide. Physicist Nigel Calder is scathing:

> "CERN has joined a long line of lesser institutions obliged to remain politically correct about the man-made global warming hypothesis. It's OK to enter "the highly political arena of the climate change debate" provided your results endorse man-made warming, but not if they support Svensmark's heresy that the Sun alters the climate by influencing the cosmic ray influx and cloud formation.

> "The once illustrious CERN laboratory ceases to be a truly scientific institute when its Director General forbids its physicists and visiting experimenters to draw the obvious scientific conclusions from their results."[102]

---

[101] http://calderup.wordpress.com/2011/07/17/%e2%80%9cno-you-mustnt-say-what-it-means%e2%80%9d/

[102] Ibid.

Svensmark's Supernova Revolution

The breaking news is Svensmark's study of supernovas. This is possibly the biggest change in our understanding of life on Earth since Darwin's theory of evolution. The CERN experiment strongly suggests that the Sun's variable magnetic activity can influence the amount of cosmic rays that hit Earth and form clouds. But what if the number of cosmic rays is itself variable? The more cosmic rays out there (regardless of the Sun's activity at the time) the more are available to hit Earth; and if there are very few, then no matter what the Sun is doing, there will be very few rays hitting Earth. Cosmic rays are given off by supernovas, massive stars that explode when they use up their fuel. So the more supernovas, the more cosmic rays.

Svensmark noticed this, and didn't stop with merely formulating his theory about cosmic rays: he analysed the entire history of supernovas in our galactic neighbourhood for the last 500 million years. What he found was a remarkable correlation between the frequency of supernovas and the diversity of marine life. Calder summarises Svensmark's results:[103]

"The long-term diversity of life in the sea depends on the sea-level set by plate tectonics and the local supernova rate set by the astrophysics, and on virtually nothing else.

"The long-term primary productivity of life in the sea – the net growth of photosynthetic microbes – depends on the supernova rate, and on virtually nothing else.

"Exceptionally close supernovae account for short-lived falls in sea-level during the past 500 million years, long-known to geophysicists but never convincingly explained..

"As the geological and astronomical records converge, the match between climate and supernova rates gets better and better, with high rates bringing icy times."

The "executive summary" of Svensmark's investigation is that all the tortuous attempts to 'explain' the Earth's long-term climate in terms of CO_2 have it completely backwards, and Svensmark has all

---

[103] Nigel Calder. http://calderup.wordpress.com/2012/04/24/a-stellar-revision-of-the-story-of-life

but proved it. When life is doing really well, as in the Carboniferous, and recently until the peak ice ages hit, it draws down the CO_2 in the atmosphere—converting it into living tissue, as anyone prior to the current period of climate insanity would have easily expected. This is why the net trend of CO_2 since the dawn of life has been downwards—we are 'using it up'!

Understanding the Supernova Theory

This section is for readers who want to follow the story in a bit more detail, but it can be safely skipped.

To start, a supernova is an exploding star. Stars come in a range of sizes. There are really small ones (brown dwarves), many small-to-medium stars (our sun is one of these), and some monsters many times the mass of our Sun. The first point to appreciate is that the bigger a star is, the shorter it lives, because it burns through all its fuel at such a fast rate that it uses it all up much quicker than a smaller star with less fuel. So, our good Sun has existed for about 4.5 billion years and has enough fuel left for another 5 billion years. A really big star, on the other hand, can burn so brightly and use its fuel so fast that its entire lifetime might be only about 10 or 15 million years. On a geological timescale, that really isn't very long at all. There have been about six 'big star' lifetimes since the end of the dinosaurs!

Stars are usually formed in batches. A 'star-forming region' consists of a huge cloud of gas and dust, and where the gas gets concentrated (perhaps by shock waves, etc., rippling through the galaxy), a patch of gas can start to contract by gravitational attraction. As it contracts, it releases energy and heats up. If there is enough of it, it will compress and heat up enough to ignite nuclear fusion—it starts to shine; it has become a star. A gas cloud will usually create dozens of stars—big ones, small ones, all sizes. And as they start to shine, they blow away the rest of the gas from which they were born.

What remains? Lots of stars, all roughly the same age, slowly drifting apart. Astronomers call this an "open cluster". The best known open cluster is easily visible to the naked eye—the Pleiades, also called the Seven Sisters. But here's the critical point for Svensmark's analysis: the big stars are the ones that end their lives as supernovas. And because these big stars live for such a short time, the supernovas all 'go off' within a few tens of millions of years of the birth of the rest of the stars in their cluster. So if we can date the smaller stars (which are still out there to be seen because small stars live a long time), we can infer the region in time and space in which those big stars would have had to 'go bang'.

By examining the open clusters in the sky today, we can work out where and when supernovas were exploding in the distant past. This is what Svensmark did for the past 500 mya, and by calculating the motion of our Sun through the galaxy, he could work out when the Sun was in a region with lots of supernovas, and when it was in regions with few or none. That told him when our solar system was bathed in lots of cosmic rays, and when there were very few. According to his cloud theory (and confirmed by the CERN experiment), that amounts to when there were lots of clouds and when there weren't.

It is important to understand that these calculations are not a 'computer model' in the same sense as the climate models which, we have already seen, have never made a correct prediction. The climate models are simulations using simplified equations, with 'parameters' (i.e. guesses) to cover the aspects of reality that the model cannot handle. On the other hand, Svensmark's calculations are simply solving the well-confirmed physical equations for how the Sun is moving through the galaxy, in order to work out where it was at various times in the past. The answers aren't exact (no calculation of a physical quantity can ever be) but the likely errors can be computed (and Svensmark computed them). Subject to those error limitations, his calculations are as good as the ones that allow us to predict the next date of an eclipse, or a coming 'king tide'. We can fly men to the Moon, rockets to Pluto, using those equations. This is proper science, not the fantasy science of the climate models.

What Does It All Mean?

Svensmark has shown that two important measures concerning life on Earth are determined more or less entirely by the rate of supernova explosions at the time (in the case of productivity of life in the sea) and by supernovas combined with the sea level (in the case of diversity of sea life). That is, conditions for life are to a very large extent driven by a factor completely beyond our control. Carbon dioxide levels are a response to these changes, not a cause of them.

We have had trouble in the past understanding these connections because the various processes operate on very different time scales. The record of Antarctic ice cores shows that CO_2 concentrations in the atmosphere *follow* temperature changes by about 800 years. This process is driven by ocean heating and cooling. Gases are less soluble in hot water than in cold, so as the oceans heat up, they drive CO_2 out of solution into the atmosphere. The turnover time of all the water in the oceans is about 800 years, which explains the time lag.

On the other hand, the Sun's activity follows a mere eleven or so year cycle, and it can be greatly different in activity from one cycle to the next. This driver modulates cosmic rays from decade to decade. But the supernova rate that determines cosmic ray abundance happens over tens of millions of years, as the massive stars in nearby newly formed open clusters reach the end of their lives and explode as supernovas. The ocean depth changes quickly when ice is being formed or is melting, but changes slowly when it is due to continental drift. All these confounding factors can easily lead the unwary to draw knee-jerk conclusions, such as accusing an effect (CO_2 concentration) of being the cause (of temperature changes, for example).

A key issue that alert readers will have spotted is that Svensmark's analysis seems to contradict much of the evidence we have looked at that shows warmth to be better for life. But this is the nature of real scientific research. It is never "settled"; the debate is never "over"; and those who claim it is are either ignorant or deceiving you. Svensmark has explained some important puzzles, but he has also opened up brand new questions that will take years to work through. We need proper scientists to do this vital work, people committed to truth, not people committed to presenting you a 'tidy' picture that is 'on message' for ideological reasons.

With that in mind, here are some points I suggest may be of importance:

- Life in the sea seems to prefer a colder temperature range than life on land. Cold seas are much more productive than tropical seas. But the same is not true of life on land; warm regions have much more life than tundras and cold mountain tops.
- Life prefers a specific range of temperatures; it can be too hot, or it can be too cold, for any given kind of life. If sea life prefers it colder than the planet would be on average, then an influence that cools things down (supernovas) will appear to be connected strongly to more life in the sea. This result might not be repeated for life on the land—and indeed, historical evidence shows pretty clearly that current temperatures are a few degrees below optimum.
- Svensmark's explanation for why cold is better for sea life is that cold stirs up more 'weather', driving more nutrients into the oceans and stirring them up better. But that is exactly what Plimer has told us: cold times have much *worse* weather (from a *land* animal's point of view!) with more terrible storms, alternating heat and cold, droughts, floods, and famines. This bad weather for land creatures stirs up the nutrients that the sea creatures feast upon.

Lastly, a thought concerning our 'programming' in the politically correct world we are all now immersed in: what is the importance of *diversity* of life? First a simple fact: the word "diversity" means "lots of different kinds"; it is a morally neutral word; but we are all programmed to uncritically accept it as a good thing. But that depends on circumstances.

When is diversity good? Well, there are lots of well-known benefits from diversity, and I need not make much of the question here. Diverse ways of looking at the world will probably throw up better new ideas when difficult problems arise than if we all thought the same. Diverse types of food makes life much more interesting and enjoyable; and the same for music, dancing, and so on. And, getting closer to our topic, diverse kinds of plants and animals means that in difficult times, some species or other will be able to take advantage of any given marginal ecological zone.

But this last fact turns diversity on its head. It might be good, but it isn't a good measure of life's well-being! When life is easy, food is plentiful, the climate balmy, and so on, almost any creature will be able to survive. What happens is that a few species do very well indeed and 'take over'. Then when times get tough, the few species find their ranges greatly reduced, and specialists with skills for dealing with the tough environments will arise to live in the difficult places. This is exactly what happened when icy conditions made life intolerable in the far north for normal brown bears—polar bears evolved with special skills to survive in cold regions.

The implication of this, since we are talking about time scales on which evolution has full scope to operate, is that times in which there is a lot of diversity of species are likely to be tough times, not easy ones.

In the final analysis then, no, CO_2 is not a major cause of planetary warming. We have identified the major drivers of Earth's climate: Our Sun's energy, modulated by cloud cover driven largely by cosmic rays. None of these factors are within human control, so let's stop trying to do a repeat performance of King Canute ordering the tides to turn back, and start dealing with whatever climate turns up. It is much easier to adapt than to try to control.

If that were the end of it, we might be tempted to say "Okay, it seems certain now that attacking carbon is completely useless, but let's do it anyway just in case." Unfortunately we know that carbon dioxide is plant food. There is 6% more greenery available now as food for both humans and wildlife thanks to industrial emissions of plant food. That reverses the entire logic of what we as a society are trying to achieve. We are, with carbon taxes and trading schemes and anti-carbon laws, working as hard as we can to starve the planet and kill people and wildlife.

6
Why Do They Demonise Carbon Dioxide?

If carbon dioxide is beneficial, if warming is on the whole good, if life on Earth flourishes in warmth and dies off in the cold, it is a perfectly natural question to ask why we are being relentlessly bombarded in the media with scare stories about global warming?

Although this is not the central concern of this book, I have found that friends with whom I discuss these questions (and I, myself, before I started looking into the subject) often find it hard to credit that so many 'authorities' would all agree with the global warming theory if it were in fact false:

- the scientific establishment (so we are told) has a 'consensus', and 'the science is settled' (although in fact over 31,000 scientists have signed a petition opposing the supposed consensus[104]);
- how could so many government agencies conspire to falsify the historical temperature record?
- how could 200,000 measurements of CO_2 concentration have been suppressed?
- why would governments only finance the science that supports the global warming theory?
- why would the IPCC (the Intergovernmental Panel on Climate Change), fabricate evidence and include non-scientific claims in its so-called scientific summaries?
- why would the world go to so much trouble setting up markets for carbon credits, and pass cap and trade laws, and so forth, if they did no good?
- why would the finance markets and the banks support the global warming theory?

---

[104] See http://www.oism.org/pproject/

- why would Big Coal support it?
- why does almost the entire mainstream media support it?
- why do pretty much all the left-wing political parties and many of the right-wing parties support it?

What I believe is happening here is the phenomenon nowadays known as the *perfect storm*. It's not a real (weather) storm, it is a lucky (or unlucky!) combination of circumstances that is just right for making some unlikely thing happen. For example, a normally safe driver might have an accident due to a 'perfect storm'. Maybe he is distracted by a spider in his car at the exact moment a child runs onto the road, causing him to lose his reaction time to brake and stop, but just then a vehicle approaching from the other direction makes it impossible for him to swerve as a backup strategy. The unlikely and unlucky combination is enough to make the accident happen, whereas if even one of the causes had not happened, or had happened a moment sooner or later, he would have had enough of a safety margin to avoid the disaster.

The global warming scare is perhaps the consummate and unluckiest perfect storm in history, a combination of accidental and deliberate factors that have mutually reinforced each other.

Let us start with the purely practical issue: money. Here is the Director of the United States' Congressional Budget Office, telling us how much money is available to you if you support the theory:

> "From 1998 through 2009, appropriations for agencies' work related to climate change totaled about $99 billion (in 2009 dollars);... During that period, the nation's commitment to climate-related technology development increased significantly, as has the forgone revenue attributable to tax preferences."[105]

American readers, that's your tax money at 'work'. But Britain and Australia, all of Europe, and the rest of the West are no better off, and tax money that could be better spent is being thrown at 'work' related to 'climate change' everywhere.

More surprisingly, 'Big Coal' is also throwing bucket-loads of cash at people prepared to work on 'green' projects. Here is Aus-

---

[105] Doug Elmendorf, Congressional Budget Office Director's Blog:
http://cboblog.cbo.gov/?p=487

tralian 'Big Coal' giving away a billion dollars:

"Australian Coal Association committed to action on Climate Change

"The coal industry wants to see Australia address climate change as part of an international solution, the Executive Director of the Australian Coal Association (ACA), Mr Ralph Hillman said today.

"The coal industry is committed to action on climate change and the industry is contributing a billion dollars over the next 10 years developing technology to cut carbon emissions from coal fired power stations by up to 90%."[106]

But surely "Big Coal" doesn't want to 'push' a theory that would make us want to give up coal?

That is exactly what they do want. Some of them, despite working in "Big Coal", are nonetheless convinced of the reality of the threat and want to 'do the right thing'. Others, more mercenary, support it for one simple reason: they know there is no 'green' alternative to fossil fuel. Wind and solar power can at most make up a small proportion, maybe ten percent, of our energy needs; meaning that the only way to make our energy usage 'clean' (according to the murderous theory that CO_2 is bad) is to "clean up" our usage of coal. But what is involved in designing "clean" coal power stations?

In short, a 'clean' coal power station is one that does not release the life-giving CO_2 plant food into the atmosphere where it can do good for humanity and for wildlife, but instead 'sequesters', or buries it, deep beneath the surface where it will languish, useless, while people and animals starve from the enfeebled ability of plants to grow strong and fruitful.

And what is the monetary cost of this anti-life policy?

In brief, it is either impossible (for no one has shown how to do it reliably yet) or it will require vast quantities of extra energy to pump that CO_2 deep into the planet. Estimates are that about an extra 40% coal usage would be needed to deny the planet the benefits of releas-

---

106 http://www.australiancoal.com.au/resources.ashx/MediaReleases/81/MediaRelease/115CBF3A363CA19CD1F054B99EDC6169/ACA_National_Release_041209_(3).pdf

ing CO_2 into the atmosphere.[107] Even the IPCC admits that carbon capture and storage (CCS) could use anywhere from 10% to 180% more energy than an equivalent power plant without CCS. And even then, not all the CO_2 would be captured.[108]

Furthermore, retrofitting existing power plants would lead to even higher costs.[109]

So, if you were 'Evil Big Coal', what would you do? If people come to believe in the global warming hoax, they might take away 10% of your trade and put it into destructive wind farms and so on, killing wildlife and ruining pristine environments, but being Evil, you don't care about that, and they will be giving you around an extra 40% (or maybe heaps more) in sales of extra coal to bury the plant food generated by the coal plants still in operation. 40% minus 10% is a 30% additional profit. That's why Big Coal supports the hoax. (And by the way, evidence is now emerging that pumping all that gas into the rock strata would likely cause earthquakes.[110])

The money motive keeps on appearing time after time. The IPCC itself was formed to address what politicians thought was a real problem; and every single person from the Head, Rajendra Pachauri,

---

[107] It's hard to pin down reliable data about the energy costs of carbon sequestration and storage, but even U.S. Government reports admit current efficiency would be reduced by up to 32%. (United States Government Accountability Office, *Coal Power Plants: Opportunities Exist for DOE to Provide Better Information on the Maturity of Key Technologies to Reduce Carbon Dioxide Emissions* http://epw.senate.gov/public/index.cfm? FuseAction=Files.View&FileStore_id=66a9e301-056d-482a-9cc6-8f1989a2675f) But such figures seldom include additional costs such as manufacturing, shipping and, in this case, finding enough places to store vast quantities of compressed gas.

[108] Pages 4-5 in IPCC, 2005: *IPCC Special Report on Carbon Dioxide Capture and Storage*. Prepared by Working Group III of the Intergovernmental Panel on Climate Change [Metz, B., O. Davidson, H. C. de Coninck, M. Loos, and L. A. Meyer (eds.)]. Cambridge University Press, Cambridge, United Kingdom and New York, NY, USA. Online at http://www.ipcc.ch/pdf/special-reports/srccs/srccs_wholereport.pdf.

[109] *Ibid* p 10.

[110] See The National Academies News: *Hydraulic Fracturing Poses Low Risk for Causing Earthquakes, But Risks Higher for Wastewater Injection Wells.* http://www8.nationalacademies.org/onpinews/newsitem.aspx?RecordID=13355

down to the office cleaners, every last one of them, depends for their pay check on the continuance of that belief. What if there were no real problem? Then no need for the IPCC! Every single academic receiving government grants for work on 'climate change' gets the money because governments and politicians, usually scientifically illiterate, are rattled and scared by the threat of horrendous damage by something they don't understand; any academic who puts up papers saying 'no problem' will likely soon be out of a job.[111]

Bret Stephens, writing in The Wall Street Journal[112], summarises just a few of the money trails that drive the climate change movement. As you read these figures, divide them by whatever amount you would consider to be a typical white-collar salary to get an idea of the numbers of people employed by the global warming scare. Then remember that every one of those employees owes their job to keeping the scare running. No scare, no salary! That's a strong incentive. Anyway, here are the figures:

- Climatic Research Unit: $19 million
- From the European Commission: almost $3 billion
- US Government:
  - $1.3 billion for NASA
  - $400 million for NOAA
  - $300 million for NSF
- California alone: $600 million
- Total estimate from HSBC Bank for a single year: $94 *billion*.

And so the money drives bad science and bad policy.

---

111 There are two shocking recent cases that appear to be exactly this: academics fired for their opposition to the climate hoax. One at UCLA:
http://www.wnd.com/2012/06/ucla-accused-of-firing-prof-for-criticizing-pollution-agenda
and one at Oregon State University:
http://wattsupwiththat.com/2012/06/11/climate-skeptic-instructor-fired-from-oregon-state-university
112 http://online.wsj.com/article/SB10001424052748703939404574566124250205490.html

Conspiracy?

Those who doubt the existence of a threat from carbon dioxide are often ridiculed as "conspiracy theorists". How can it possibly be, many ask, that the entire scientific and political establishments are in on some massive secret plot to demonise carbon? One thing that needs to be made crystal clear is that there is *no need* to invoke the idea of a secret global conspiracy to explain what is happening. Ordinary human fallibility coupled with ordinary human needs to get ahead in life, put bread on the table, make a name for oneself, and so on, suffice to explain a great deal of it.

Having said that, it is nonetheless true that some people are conspiring on schemes connected with global warming. Human beings *do* conduct conspiracies! Criminals conspire to rob banks, politicians conspire to spin stories to support their cause, market traders conspire to create market conditions that will make their trades pay off, people running a company will conspire to make their products attractive in the market. Not all these conspiracies are big ones (though some are) and not all are illegal (though some are). So it seems strange that some find it hard to believe that powerful movers and shakers will conspire from time to time, and when they do their conspiracies might be big ones—after all, they have the power and money to finance them.

So, although there is no need to suspect a single huge, dark, worldwide conspiracy, we should not be surprised if some dealings of a conspiratorial nature happen now and again in this connection. And, just as with a huge bank robbery (only worse), we should not be surprised if the conspiracies that do exist have big effects that cause a lot of damage.

ClimateGate

The big news in 2009, just as the establishment was gearing up for the big Copenhagen conference, a scandal broke. A whistleblower released a collection of emails and other related material from the University of East Anglia. It would appear that the collection had been made in response to freedom of information (FOI) requests in case the University was forced to comply and release the material. But whatever its purpose, it blew apart any possibility of a treaty arising in Copenhagen.

That might be just as well. Lord Christopher Monckton of Brenchley writes that the secret purpose of the treaty was:

> "... to discuss a treaty to inflict an unelected and tyrannical global government on us, with vast and unprecedented powers to control all once-free world markets and to tax and regulate the world's wealthier nations for its own enrichment: in short, to bring freedom, democracy, and prosperity to an instant end worldwide, at the stroke of a pen, on the pretext of addressing what is now known to be the non-problem of manmade "global warming"."[113]

Has the good Lord lost his marbles with talk about tyranny and world government? Perhaps your first reaction is "He surely must have, that's tinfoil hat stuff!" Luckily there are already other writers who have taken it upon themselves to investigate it in depth; Monckton's paper referenced above is a good place to start. But if we have been lied to about the basic science—and we have! —then why shouldn't the politically motivated movers and shakers behind that lie not also lie to us about other things?[114]

The Motivation?

Motives, of course, are unique to each individual. One person joins a physics department, say, out of curiosity, another for the pay cheque, another for the glory of doing something big in science. It has to be a similar thing with the players in the warming hoax.

But having said that, I believe the basic motivation, the one that

---

113 Christopher Monckton of Brenchley. Caught Green-handed! Cold facts about the hot topic of global temperature change after the Climategate scandal. http://scienceandpublicpolicy.org/images/stories/papers/originals/Monckton-Caught%20Green-Handed%20Climategate%20Scandal.pdf

114 A lively book that looks at the more political aspects of the whole fiasco is Delingpole's *Killing the Earth to Save It* (Australian title), also published as *Watermelons—the green movement's true colors.*

activated the original players who got the whole ball rolling, was a similar one for most of the key movers and shakers: a combination of damaging ideological beliefs with a misguided way of 'seeing' the natural world. This needs some explanation—but first, a warning:

We all know cases of people who have some self-harming habits or beliefs but who, by virtue of being 'too close' to themselves, can't see the wood for the trees. It is very difficult indeed to critique ourselves. What I am going to suggest in the following is a self-critique of all of us: the entire human race. And since the subject of discussion is ourselves in all our various roles and from all our different perspectives, it will be very hard for me to say it without many who are reading this seeing it as a criticism. If so, apologies, but it is a criticism of us all, myself included.

Things are going very wrong on our planet, and most of us admit that humans are involved in causing a lot of it, so a self-critique might hurt. But unless we know the truth, we will never fix things, so the choice is between shielding ourselves from pain by *pretending* we are fixing things even when we are not, or suffering a bit of psychological pain and *actually fixing* the problems.

With that in mind, let me suggest an explanation for our difficulties. Human beings are one species on a wonderful planet—a very remarkable species, to be sure: the only one capable of fully symbolic language, of incredible levels of mental abstraction, and the toolmaker supreme— but nevertheless, a species which, like all others, has instincts, behavioural patterns, comfort zones, likes and dislikes.

We also share many qualities with many other higher animals. I have personally witnessed an Australian magpie form a theory of mind, that is, the bird deduced the contents of another animal's mind (namely mine); it also calculated how to put into that mind a specific piece of abstract information which, obviously, it could not tell to me in words. [115]

That some other animals can also master higher mental functions such as this is usually denied by scientists and other intelligentsia,

---

[115] For a real-life example that happened to me, see
http://wingedhearts.org/stories/birds-can-communicate-without-words

who tend to dismiss all personal observation as mere 'anecdotal' evidence.[116] I believe that this way of looking at what counts as evidence in science is itself severely misguided, but it is not the point I am trying to make here.

My point is that we are animals like many others— but that fact is not a put-down because there is nothing insignificant or trivial about any creature with a brain that can form beliefs, be aware, love or hate, form desires, protect its young, and do any of hundreds of things that many higher animals do, but which we often foolishly think can only be done by that wonder species, *homo sapiens sapiens*. We may be exceptional in many ways, but if we believe, as we often do, that we are so special we are utterly unlike every other species on Earth, then, tragically, many things go wrong.

The Garden of Eden story from the Book of Genesis is a profound analysis of the separation, the sense of loss, of exile, of deep primeval sadness, which underlies human feelings about our relations with other species and our place in the world. Once having conceived the unutterable vastness of the separation between our capacities and those of other species, we almost necessarily cease to think of ourselves as part of nature: 'nature' is the alien, but mysteriously alluring, place 'out there', whilst 'humanity' is the civilised, industrialised, sanitised, sterilised world 'in here'.

So we each, to some extent, carry within ourselves the paradox of both loving our world and the other beautiful life forms which share it with us, and feeling our separation and our exile from its beauty and wholeness. Are we not all entranced to see on the internet a video of a wild adult lion rushing up to and embracing as friends the humans who brought it up as a cub! How much we are gladdened to watch a wildlife rescue show on television and see an animal saved by wildlife carers! But most of us cannot go there and do those wild

---

116 There are some signs that things might be changing. In story after story, year after year, a consistent picture has emerged from the witnesses of animal lovers and careful observers all over the planet. And now the reality of animal consciousness has started to be recognised by 'real' science—a major change of mind formalised in the The Cambridge Declaration on Consciousness (see http://fcmconference.org/-img/CambridgeDeclarationOnConsciousness.pdf).

things: we know that tomorrow, we will be back at work, maybe in an office, or if in the outdoors, at least in a 'tamed' part of the environment such as a farm or some other humanised place. How the separation tortures us! Yes, the level of torture is low, perhaps below the threshold of consciousness, but we are subjected to this separation, this alienation from the natural world, for our entire lives.

So what happens? In various ways we have to 'deal' with it. Some shut it out and revel in the superiority of humanity over nature, despising the 'bleeding hearts' who would worry about a few birds and animals when a new freeway has to be built. Others, some very few others, do escape back to nature and turn their backs on human folly, as they see it.[117] Some just 'live with' or ignore the buried pain. But others see the issue in moral terms, terms in which humans are the devils and the natural world the angels.

Of course there are a thousand variations on this theme, and no picture I could paint would be completely accurate about any particular one of us, but this is in general terms what I believe happens: We sense a separation, and we see our own power as a species and our complete subjugation of the world of nature. Then when we see how our great powers often harm the natural world, we develop a sense of guilt; then comes the need for absolution, by which we free ourselves of our guilt. From that comes the need for atonement, by which we put right the wrong by paying for it somehow. From then onwards, all our thoughts and opinions about the environment are no longer controlled by logic, nor even by our original genuine love for the natural world, but instead are ruled by our deep inner pain and a resulting selfishness masquerading as love: namely that our actions should be, not what really benefits our planet and the real living creatures who deserve our help, but instead whatever best alleviates the pain in our souls.

Here is one concrete example. Under the new Australian "carbon" tax, the price of refrigerant R404A will go from \$30/kg to

---

[117] This is the option often considered the noblest; but if we all did it, the planet could not support 10% of the current human population, and even that amount would have to wreck the environment to survive as hunter-gatherers.

$318/kg, a 1,060% increase![118] This will make any fridge or air conditioner that goes wrong and needs 'topping up' to be 'not worth repairing'. Brand new equipment made overseas where the refrigerant can be obtained at the low cost will be imported and the old, fully serviceable equipment will go on the scrap heap. This has all of the following consequences:

- the old equipment adds to land fill and pollution problems (meaning *real* pollution, not beneficial CO_2);
- all the raw materials for the new equipment need to be dug up, transported (multiple times), processed, manufactured into components;
- the new equipment needs to be built and transported, then installed in place of the old.

Needless to say, all this effort wastes huge amounts of non-renewable oil, coal and electricity, all of which puts more CO_2 into the atmosphere. (Yes, I hope by now we all agree that is in fact a benefit, but the carbon tax is based on the assumption that is is harmful, so it fails by its own standards.)

So in the end, the tax on refrigerant will save not one single ounce of gas because, unless we all go so broke we can't even afford a fridge, well, we'll all still have fridges and the only question is whether we repair the old one or waste non-renewables making a new one. And the logical, loving, genuinely caring option is to top up the old one; but the feel-good "we're reducing our carbon footprint" eco-solution is to damage the planet making an entire new fridge. That is just one example of how we are now being forced by law to do feel-good but damaging things instead of simply following the cheap and environmentally safe path.

Returning to our main theme, why we are so often led to do feel-good harm to nature rather than what actually works, one might think that actually helping wild creatures to live happier lives would alleviate the pain that arises from our guilt at harming the natural

---

118 http://www.redmeatinnovation.com.au/project-reports/report-categories/professional-development/ampc-advancing-the-future-conference-june-2011/day-2-m40-refrigeration-optimisation

world, but not necessarily. Guilt and its expiation psychologically requires action. But in the global warming case, *inaction* is the best way to help the wild world. CO_2 is not harming the planet, in fact our industrial emissions of it have fed hundreds of millions of extra people and countless wildlife. Simply going on enjoying the fruit of abundant power and technology, and also helping the poorer nations rise to our own standards and enjoy life along with us, that is the best way to increase CO_2 and feed an extra billion people over the next forty years, and thereby reduce population pressure to convert wild areas to human food production.

But enjoying oneself cannot expiate guilt. Only stopping driving a car, turning off an air conditioner and either boiling or freezing, foregoing safe but slightly less efficient light bulbs in favour of dangerous mercury-filled contraptions that will spread toxicity around the landscape—only by suffering in such ways as these (or, in the case of many activists, forcing someone else to suffer in these ways) can one *feel* the atonement and regain innocence. The price of being different from all other species, whether that difference is real (as in the case of our capacity for grammatical language) or imaginary (as in the case of claims that animals have no emotions), is to feel exiled, and the consequence of that is doing things that don't really make sense if our goal truly were to dispassionately help humanity and help the environment. Princeton physicist and climate realist William Happer offers a slightly unkind, but quite accurate, comment on the problem:

> "There are people who just need a cause that's bigger than themselves, then they can feel virtuous and say other people are not virtuous."[119]

Is this relevant to the science of 'climate change'? Whether my thoughts about human guilt are right or not, we cannot deny that humans have an in-built tendency to think of themselves as sinful, fallen, a blot on the planet; even, in the words of Prince Phillip, a plague. Paradoxically, one feels virtuous by admitting one's sins.

---

[119] Quoted in *Climatologists are no Einsteins, says his successor,* by Paul Mulshine. See http://blog.nj.com/njv_paul_mulshine/2013/04/climatologists_are_no-_einstein.html

That is the inner logic behind all the public confessions seen, for example, during the Chinese cultural revolution, and of sinning celebrities in western society such as Tiger Woods. We have a deep need, not to actually *be* sinless (to use a biblical word), but to gain the high *moral status* of a repentant sinner. We have to be seen and admired for confessing our own sins and atoning for them. Human damage to the environment, directly tapping into our guilt and separation anxieties, provides the consummate rationalisation for dysfunctional ways to satisfy these inner needs.

We can gain further insight into this problem from this significant passage:

> "In searching for a common enemy against whom we can unite, we came up with the idea that pollution, the threat of global warming, water shortages, famine and the like, would fit the bill. In their totality and their interactions these phenomena do constitute a common threat which must be confronted by everyone together. But in designating these dangers as the enemy, we fall into the trap [of] mistaking symptoms for causes. All these dangers are caused by *human* intervention in natural processes, and it is only through changed attitudes and behaviour that they can be overcome. The real enemy then is humanity itself."[120] (italics in original)

This passage does not come from a single individual, who might be thought to be unbalanced, however powerful he or she might be; it comes from the Club of Rome, since 1968 one of the most influential of all think-tanks, consultant to the United Nations, and composed of a great many of the most powerful people on Earth. This passage states in black and white that the whole global warming scare was selected opportunistically for an ulterior motive; but my main reason for including it is the glimpse of the human psyche that it affords us.

In the first place, this passage reflects human qualities that are *admirable*: unity, implied friendships and cooperation, and so on; it is very unjust indeed to imagine that the original impulse behind the alarmism was malevolent, no matter how destructive it has become.

---

120 Alexander King and Bertrand Schneider, The First Global Revolution; a Report by the Council of the Club of Rome. p 75.
Online at http://www.archive.org/details/TheFirstGlobalRevolution.

But do we not also see in this passage a segue, an unconscious transition from this positive attitude about human nature towards the same alienation and guilt I discussed above? At "mistaking symptoms for causes" we have a reasonable warning against a common mistake, but what follows this in the text does not follow logically, and the size of the error grows exponentially in the few words remaining to the end of the paragraph. No, we do not have any evidence that global warming is a human-caused dangerous problem, none whatever. In fact it alleviates another problem they mention, famine, by lengthening growing seasons and making more of the Earth habitable. And then the collective 'auto-immune disease' of the human psyche kicks in and we are given the startling non sequitur: *the real enemy is humanity itself*. The alienated human mind turns upon itself.

Perhaps we could place a more positive construction upon this passage: maybe it is only saying that we are messing ourselves up? That might make sense if this were the only passage with a negative interpretation, but it is one of hundreds by some of the most respected environmentalists in the world, as we shall see in a later chapter.[121] But whatever the intention, we can surely ask: why did the passage not end on the positive note of unity, friendship, and co-operation, rather than moving on to the negative and oppositional point of casting 'humanity' as the enemy of 'us'?

This is not a trivial question: it cannot have escaped anyone's notice that if you turn on the television and watch almost any nature documentary made in the past twenty years, there is a standard 'script': tell us about the cute, cuddly, or magnificent wild animals, then tell us how they are threatened, then finally say that it is all our fault. At that point, where we are made to feel ashamed and guilty, most of these shows stop; they seldom proceed further to telling us how we can actually do something positive—unless it is the completely wrong advice to 'cut our carbon footprint'. It is hard to escape the conclusion that the shame and guilt is the unconscious objective rather than the fixing of the problem.

---

121 A collection of some of the most shocking can be found at:
http://green-agenda.com/index.html.

Carbon is Life

Humans frequently harm the environment—that's as plain a fact as anything can be, so it is exceedingly hard to argue against human guilt when it is expressed inappropriately on environmental subjects. Unfortunately not only do humans harm the environment, but they must *believe* this even when it isn't true in order to provide an outlet for atonement of guilt; so telling a committed environmentalist that — good news!— global warming is minuscule, harmless or beneficial, and not being caused by human industrial activity, is not only pointless, it is a direct attack, a threat against the core psychic identity of the human being as exile from the paradise of nature and (repentant) sinner. The matter ceases to be a question for scientific investigation and becomes a deep moral and religious commitment. Once someone has gone down this path, to point out to them the facts about the value of carbon dioxide to life is not to discuss science, it is to assume the form of the devil attempting to invade their soul.

But things don't stop there. Each person who allows his- or herself to be 'captured' by their inner sense of alienation usually comes to the global warming question with a whole raft of other commitments that were motivated by the same psychic pain. Often but not always, these take the form of political commitments, usually to parties or ideologies opposed to 'materialism', 'capitalism', and so on.

And disliking capitalism isn't hard! Some capitalists, such as Hayek, support it simply because they accept the value of a free market in creating wealth and the impossibility of creating it in a command economy. But they are outnumbered (surely?) by capitalists seeing the system as a way to appropriate wealth for themselves. We all know of multinationals wrecking environments, ruining lives, doing cost-benefit analyses on questions like whether to install a safety system that would prevent needless deaths, and so on.

The dysfunctional ones on the side of capitalism are also fleeing from the human condition of exile, but instead of trying to atone for something beyond their control, they have buried the knowledge of it deep within their unconscious. They busy themselves with money, getting ahead, winning power and influence, and so on. They hurl contempt at their buried pain, as if to defeat it rather than accommodate it.

Like the rest of us, these people, too, are not psychically whole. No better illustration than this is the case of so many American lawyers on an hourly salary that ordinary people would consider obscene, working themselves into the ground doing eighty-hour weeks or more. What for? It is no wonder that environmentalists, thinking themselves to be virtuous defenders of the planet, have little but contempt for their opposite numbers on the capitalist side of the divide.

And yet both the greens and the capitalists are harming the planet as a result of the very same alienation. Whether one takes a 'religious' message from it or not, the story of Adam and Eve's expulsion from innocence is profound and too often ignored as silly or outdated.

The problem here, the reason global warming became a 'perfect storm' and created a worldwide panic over a non-problem, is that almost all of us, no matter which way we have personally chosen to ameliorate our inner psychic guilt, can find a way to use global warming for our own purposes:

- environmentalists can feel virtuous for their 'sacrifice at the altar' as they "reduce their carbon footprint" and force others to do likewise;
- unprincipled money- and power-brokers can make money and gain power easily because the 'commodity' being manipulated is entirely fictitious (*not* emitting some CO_2 — European regulators estimate that some 90% of trades on the carbon market are fraudulent);
- now that "reducing your carbon footprint" has become a point of honour (even though, as we have seen, it is actually harming the planet), everyone from ordinary people to politicians to business people can gain esteem and moral kudos by supporting the hoax;
- academics gain grants and promotions by inventing research that has the desired (pro-hoax) conclusions;
- Big Coal can sell at least 30% more coal to power 'carbon se-

questration' in a world that is in panic about carbon dioxide;

- all the people employed to 'solve' the non-problem (such as the members of the IPCC) depend on the hoax for their livelihood.

And so, by some combination of virtue, guilt, avarice, desire for money or power, religious or political belief, almost everyone has some vested interest in the continuance of the mass delusion. We hate the very food that grows the plants that we and our fellow non-human animals all eat, we hate the molecule that is the foundation of life on earth, and we hate the key precious element, carbon, that makes life possible at all. We seem to be locked in. Yet we had better break out, because, as an unknown but exceedingly wise ancient once said, "Those whom the gods would destroy, they first make mad."

Our Prospects

It is close to unbelievable how many, and how powerful, are the people pushing global warming for reasons unconnected to scientific truth. Timothy Wirth, later Bill Clinton's Under Secretary of State for Global Affairs is quoted as saying "We've got to ride the global warming issue. Even if the theory of global warming is wrong we will be doing the right thing, in terms of economic policy and environmental policy." And Christine Stewart, former Canadian Minister of the Environment is reported as saying "No matter if the science of global warming is all phony ... climate change provides the greatest opportunity to bring about justice and equality in the world." [122] These examples could be multiplied endlessly. The point is that *when things are done for an unspoken reason, then being rational about the spoken one won't result in change, even if you win the argument.* The primary motivations here are, as we have seen, unrelated to the sci-

---

[122] Quoted by Berit Kyos in http://www.newswithviews.com/BeritKjos/kjos76.htm as follows: Christine Stewart, speaking before editors of the Calgary Herald, 1998. Quoted by Terence Corcoran, "Global Warming: The Real Agenda," Financial Post, 26 December 1998, from the Calgary Herald, December, 14, 1998. Cited by Fred Singer, page 4.

ence of the atmosphere or its genuine relationship to human activity.

So what, specifically, can we expect from the global warming scare?

- ruinous taxes and charges will be introduced to fight a nonexistent problem;
- you will be 'trained' to accept hardship—your compassion will be misused to further a political agenda;
- you will be forced to acquiesce as more and more controls are placed upon your life;
- you will acquiesce as your living standard drops and costs increase (for example the retiring leader of Greenpeace is reported to have said that there is an urgent need for the suppression of economic growth in the United States and around the world[123]);
- while many of the 'ringleaders' are on record as hating humanity even to the point of wanting to exterminate us, and while they themselves live by capitalist standards (fabulously rich in many cases), they will train you to be a docile 'unselfish' socialist
- the power of unelected UN and other international bodies will grow so big that they become unstoppable; democracy will cease to function effectively and our lives will pass out of our own control.

Yes, these harmful developments will have to be fought on the public stage and by putting forth candidates in elections and lobbying politicians and opposing carbon reduction schemes and so forth; but the inner psychic damage within us all is the real problem, and helping and nurturing others so as to heal the archetypal pain in ourselves and in others, *that* is the real battle and it requires love and friendship, not hostility.

---

123 http://wattsupwiththat.com/2009/08/19/ice-capades-greenpeace-recants-polar-ice-claim
http://news.bbc.co.uk/2/hi/programmes/hardtalk/8184392.stm

7

What Is the Real Climate Crisis?

There does not, of course, have to *be* a real climate crisis. The entire thing could simply be a beat-up best forgotten. But as it happens, there are some dangers that we would do well to think about.

A Killer Ice Age

Detailed historical records exist going back thousands of years that show clearly that warm times foster prosperity, whilst cold times are host to war, famine, disease, untimely death, and general misery.

There is evidence that between seventy-one and seventy thousand years ago, during the last ice age, a bitterly cold period brought the human species to the edge of extinction. Our ancestors suffered what is now called a 'population bottleneck': their numbers fell so low as to reduce the genetic variability of our species to dangerous levels. Something similar, but worse, happened at one time to ancestors of the modern cheetah. Cheetahs are almost identical genetically, which means that, for example, if one becomes susceptible to a deadly disease, all of them are likely in danger from it. Looking about us at the wonderful variation human beings are capable of, it seems hard to believe we nearly suffered such a fate.

Estimates vary as to how many humans were alive during the bottleneck. Some authors say 10,000, others as few as forty individuals.[124] That was the closest we came to obliteration, and it was deadly

---

124 Stanley H. Ambrose, Late Pleistocene human population bottlenecks, volcanic winter, and differentiation of modern humans. *Journal of Human Evolution* (1998) 34, 623–651. But note that some authors deny the existence of the bottleneck, for example John Hawks, Keith Hunley, Sang-Hee Lee, and Milford Wolpoff, Population Bottlenecks and Pleistocene Human Evolution. *Mol Biol Evol* (2000)

cold that did it, not heat.

Some thousands of years later in that same ice age, our human cousins the Neandertals, with brains larger than our own, were wiped out. This is all the more shocking when we remember that the Neandertals were a human species specially adapted for cold, and yet cold got them all, whereas they survived and prospered during at least one previous warm interglacial, the Eemian, which, as we saw earlier, was warmer than the current Holocene interglacial.

Ice ages are a rarity in the long-term history of our planet, but when they happen, they spell disaster for life on land—and extreme ice ages spell disaster for everybody. Warm periods, on the other hand, are the usual state of affairs and life commonly prospers during them.

As we have seen, the present warm spell is called an interglacial because it is, on a geological timescale, a brief interlude between ice ages. How do we know another ice age is coming? Because, for the past 2.5 million years, ice ages have been following one another, with brief respites, at almost regular intervals. They are happening now on approximately a 100,000 year cycle and another one is due. We know this because ice ages roughly follow changes in the Earth's orbit known as Milankovitch cycles. These orbital changes are locked in and are inevitable, barring some much worse disaster such as a major meteorite strike.

A British newspaper, which had better remain nameless, published an article showing illustrations of the 'terrible fate' awaiting England as the alleged global warming gets under way: a climate like Spain's! One cannot help but wonder why so many Brits retire to just that country when they no longer have to do the daily grind in London or Manchester or some other British city of presumably delightful climate? Perhaps that newspaper should have published illustrations of the *real* danger: London buried under a kilometre-thick glacier. That is exactly what really did happen during the past ice age. The NASA image shown in Figure 26 shows winter snow over North America.

17(1): 2-22.

Figure 26: North American Winter Snow

Each winter vast tracts of our planet are rendered close to uninhabitable; hibernators, migratory birds, and a few hardy specialists can live there—but in almost all cases only because of the life-giving return of the summer warmth. This yearly advance of ice is not normal in the Earth's geological history, it is a life-threatening abnormality, a planetary disease, notwithstanding the fact that a few hardy specialists such as polar bears and penguins have become adapted to it. Now imagine the snow that we see in this satellite image never melting, instead accumulating year after year, turning to ice, getting ever thicker and more extensive for 90,000 years. That is the reality of the recurring ice ages. For 2.5 million years, our planet has had a severe and dangerous chill, not a fever, and any human-caused warming that we can manage to contribute, like a good warming broth for a sick person, is a blessing.

This can be seen most clearly in a temperature graph. Figure 27 shows temperatures of the past 400,000 years as reconstructed from ice cores taken from Antarctica.

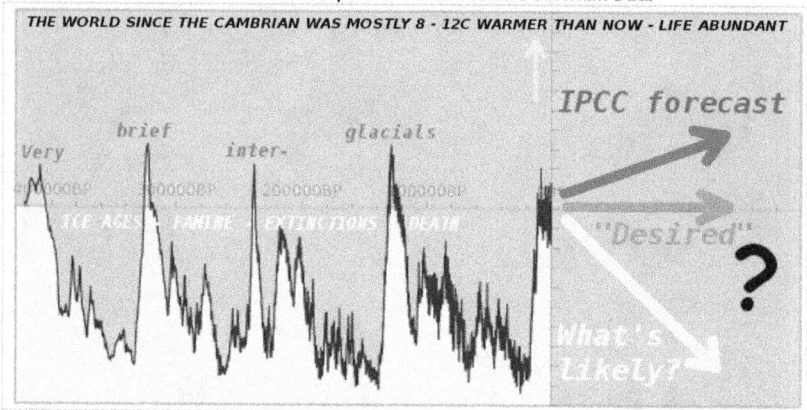

Figure 27: The past four of dozens of repeating ice-age cycles over the last 2.5my.

The repeating pattern shown here has been going on, as noted above, for 2.5 million years, getting successively colder cycle after cycle. The low points in this graph are the times when life has had to struggle to survive—or go under, as the Neandertals did. The high points are times when life was abundant and species proliferated. The previous interglacial, the Eemian, was warmer than now and teemed with life—no "global warming" problem whatever.

The Death of Everything

As many alarmists such as Al Gore point out, the CO_2 concentration plot over this time looks almost like a copy of the temperature plot. What they usually fail to tell us, though, is that the CO_2 changes *follow* the temperature changes (lagging by about 800 years), not the other way around, so temperature is the cause, whilst changes in CO_2 are the effect. This fact alone blows the anthropogenic global warming theory out of the water. But that is not the worst of it.

Because CO_2 concentration is following temperature, as successive ice ages get ever colder, the CO_2 concentration gets ever lower. The pre-industrial value for CO_2 concentration (prior to large-scale

Carbon is Life

human emissions) was about 280ppm (parts per million) and now, thanks to human contributions, it has risen to over 390ppm. But at the depth of the last ice age, the concentration fell to as little as 180ppm.

But let us remember that CO_2 is *the* critical chemical for photosynthesis in plants. As a result of human CO_2 contributions, the world is now steadily getting greener. But what about the other direction? At a partial CO_2 pressure of 15 Pascals, photosynthesis by most[125] plants (95% of all plant species) will cease[126]. That translates to 150ppm. In other words, at the depth of the last ice age, the planet came to within a whisker, a mere 30ppm, of not having enough carbon dioxide to keep almost all plants from dying! Only plants that use a very special method of photosynthesis (known as the C4 pathway) would have escaped, and even their growth would have been drastically slowed, with terrible consequences for the planet's ability to feed the world's herbivorous animals. The C4 plants are mostly grasses. Every herbivore that couldn't eat a "C4" plant (i.e. mainly grass) would have gone extinct, followed by all the carnivores that depend on them.

Could the world survive the end of 95% of all plant species and the animals that depended on them, directly or indirectly? Almost certainly not, in its current form—it would be the end of life *as we know it*. As for the next ice age? Remember they have been getting successively colder, and because CO_2 concentration depends on temperature and not the other way around, the concentration of CO_2 would be expected to fall even lower next time.

There are thousands of species of animals that cannot eat grass and rely on other plants for their only food source. Even if the planet were lucky enough that some life survived, its biodiversity

---

125 These are plants using what is technically called the C3 method of photosynthesis.

126 King-Fai Li1, Kaveh Pahlevan, Joseph L. Kirschvink, and Yuk L. Yung. *Atmospheric pressure as a natural climate regulator for a terrestrial planet with a biosphere.* PNAS 2009 106 (24) 9576-9579; published ahead of print June 1, 2009, doi:10.1073/pnas.0809436106

would be smashed, limited almost entirely to grass, grass eaters, and eaters of grass eaters. As the biosphere experiment proved (in which about the only thing that flourished was cockroaches[127]), an ecosystem that lacks biodiversity can fail for all sorts of reasons that we are not yet capable of analysing or predicting.[128]

But for the good news, the one redeeming factor: even if the deadly cold returns, as long as humans keep on emitting CO_2, we should be safe from the cataclysm of the world's plants not having enough nutrient to grow and create food. One must wonder, though, why the warming alarmists never seem to mention the critical importance of carbon dioxide in the food chain?

Have We Already Saved Earth From Icy Disaster?

Once we cease to believe the myths that the global warming scaremongers have carefully constructed all around us, and we see that a bit more warmth would on the whole be good, we might ask if we have done enough to warm the planet and ward off the next ice age. In other words, maybe the presence of major industrial societies might have changed the world for the better. And indeed, once we let go of our obsession with villainising the plant nutrient carbon dioxide, we can see that there is indeed a way in which humans cause global warming.

Put simply, our industrial civilisation gets its 'muscle' by using machines to do what had to be done by human or animal power in centuries past. All machines, including biological machines such as our own bodies, work by converting fuel into useful work.

But industrial civilisation does a lot more useful work than raw human or animal power could ever do in times past. A single person driving a modern air-conditioned tractor can harvest a field faster

---

[127] http://ksjtracker.mit.edu/2011/04/27/biosphere-2-turns-20-still-seeking-a-reason-to-exist

[128] Ironically, one reason for failure seems to have been that the CO_2 in the biosphere reacted with the concrete and was thus pulled out of the atmosphere before it could be used in photosynthesis. http://biology.kenyon.edu/slonc/bio3/-2000projects/carroll_d_walker_e/whatwentwrong.html

than a small army of farm labourers in, say, the eighteenth century. Each modern westerner routinely commands forces and does tasks far beyond unaided human capacity, whether driving a car (or even taking a train, if you are energy-conscious), storing food in a refrigerator, or, in fact, using almost any device at all.

According to the laws of thermodynamics, doing useful work must necessarily lose some of the energy from the fuel in the form of waste heat; and that heat, well, heats. In other words, because of all the extra useful work we do, we create excess heat that would not have been here otherwise, and that heat has to be dissipated somehow or it will raise the temperature.

The factors that have caused the ice ages, as we saw, seem to be related to small changes in insolation (heating) by the Sun. The changes can happen because the Sun's energy output changes or because of cyclic changes in the Earth's orbit and inclination, etc., changing the amount of heat that arrives on the surface. Changes in the Earth's orbit are believed to be the triggers for the onset of ice ages, and the changes in heating caused by those changes are thought to be quite small compared to the total power output of the Sun. This might lead us to suspect that human-caused changes in the amount of heat at the surface might indeed have a significant effect on the climate.

To answer this question, we need to compare the amount of variation due to the Sun with the amount of heat emitted by industrial civilisation. if the latter is 'in the same ballpark' as the former, then human civilisation might be holding off the onset of a new ice age.

How does it work? One theory says the driver is Northern Hemisphere summer cooling.[129] At our current stage in geological history, the North Pole is surrounded by land masses, which are snowed under every winter. If the summers became just a bit colder, then some of that winter snow would remain on the ground throughout summer, and would then turn to ice. Ice reflects sunlight much better than green plants or dirt or even liquid water, so the

---

129 See for example: http://oceanworld.tamu.edu/students/iceage/iceage2.htm.

cooling would accelerate and the next summer would be even colder and leave even more ice lying around. And so the planet falls into an ice age. Retained heat in the oceans slows down the changes and 'smooths over' short-term effects, but once the process starts, the killing ice eventually reclaims its deathly kingdom.

Dr David Archibald suggests[130] that a key measure of this process is the amount of insolation at 65° north latitude. The power of the Sun at 65°N is about 476 Watts per square metre. That means that at midday in mid-summer at, say, Reykjavik (at 64°N, almost the only significant city anywhere close to 65°N), the Sun has about the power of five old-style incandescent light bulbs. When the summer sun at this latitude is sufficient to melt the winter snowfall, all is well.

Other factors in this calculation are the length of summer (because, for example, a longer, but slightly cooler summer might melt more ice than a shorter warmer one) and how high in the sky the Sun is in mid-summer. And the higher it is in summer, the deeper and colder the long winter 'night' will be. The factors are complex and researchers disagree as to how exactly they should be combined in order to make good predictions, but some combination of these factors apparently decides whether we bask in life-giving warmth or flee the deadly cold. We cannot hope to make predictions from the kind of short overview we are doing here, but we can get an idea of the magnitudes involved.

How much radiant energy the Sun has in the past or will in the future shine upon the Earth at this latitude can be reliably calculated from basic physical and astronomical properties of the way the Earth orbits the Sun and how that orbit changes with time. This is not an uncertain thing like the forecasts of climate models; it is not easy to calculate, but it depends only upon the extremely well verified equations of Newtonian physics (or, if you prefer a few thousandths of a percent more accuracy, relativity). If we didn't know how to do these calculations, we could never have landed on the Moon or flown

---

[130] Dr David Archibald at http://wattsupwiththat.com/2009/02/23/ice-ages-and-sea-level/

discovery missions past Saturn and on to Uranus and Neptune. Yes, we *do* know how to make these calculations and we know it very reliably.

When the calculations are done, we find that at the depth of the last ice age, around 22,000 years ago, the Sun's power (again at 65°N) was around 463Wm^{-2}. On the other hand, at the height of our own interglacial, the Holocene, which occurred about 11,000 years ago (yes, we have been on the downward slope ever since—though you would never guess it from the hairy scary stories about warming in the media) the summer insolation at 65°N was about 527Wm^{-2}. In other words, we have:

What	When	Sun's Power
Previous Ice Age	22,000 years ago	463Wm^{-2}
Holocene Peak	11,000 years ago	527Wm^{-2}
The "Perfect Climate"[131]	Now	476Wm^{-2}

From these figures, we may make the following inferences:

- The difference between peak warmth and deepest cold was around 55Wm^{-2};
- The current value, being only 13Wm^{-2} above the value at the depth of the ice age, is almost all the way back to 'cold conditions';

How do these power figures compare with human energy output (mainly by burning fossil fuels)?

Human energy usage in 2006 was 491 EJ[132]. This translates to an average power usage of 15.56 terawatts[133] (power is energy expended per second, so we divide by the number of seconds in a year to convert yearly energy to power). To compare this with the Sun's power as discussed above, we need to average this over the entire planet. The Earth's surface area is 510 million sq. km., which gives 30,500 W per sq. km, or 0.03Wm^{-2}. One final adjustment is needed to

---

131 That is, if you believe the media scare stories about a few hundredths of a degree of warming.

132 http://www.world-nuclear.org/info/inf16.html. EJ equals exajoule, or 10^{18} joules.

133 A Watt is a joule per second, and a terawatt is 10^{12} Watts.

allow us to do the comparison: the Sun's insolation given above was as received at noon, whereas this figure is an average over the whole planet. Since the planet's area is four times the areas of a circle of the same radius, we must multiply by four, giving about $0.12Wm^{-2}$ as our final figure for comparison.

The human energy output of about $0.12Wm^{-2}$ is clearly overpowered by even the smallest of the numbers we have looked at so far. The $13Wm^{-2}$ difference between ice age conditions and today is at least a hundred times larger than human energy output. We might delay a killer ice age slightly, but our heating of the planet is nowhere near large enough to save us in the long term.

And We Face Many Other Deadly Threats

The bad news is that we don't need a full-scale plunge into the next ice age to cause mass death and misery. Since the Little Ice Age ended around 1850, the population has gone from a little over one billion to seven billion. If the planet's food-producing systems were reduced by cold to their capacities during the Little Ice Age, they would be incapable of sustaining the current population.

The rapid population growth over the past 160 years has happened, not only because the world has become more generous due to warmth and increased CO_2 but also because we have learned how to make more food from the same resources: better cropping methods, transportation, planning, engineering, and so on, allowing us to do far better than our ancestors did—modern technology is the prime support for our increased ability to feed the world.

This is a key reason why the emotional rejection of modern technology as 'unnatural' is so very dangerous. From time to time the media features some affluent technologically enriched westerner—American, Australian, Canadian, or whatever—who decides to 'go back to nature', lug their own water, farm their own crop, ride a bicycle, and so on. This is usually reported favourably by careless mainstream journalists: "getting back to a wholesome lifestyle", "getting in touch with nature", and so on. Comments about humans as a plague on the planet with their all-devouring, sterile technology are usually not too far behind. Indeed, this was the theme of the wildly popular movie, Avatar.

Now Avatar was a fun movie, for sure, but I have a question for its creator James Cameron: Let's imagine that there are two killer asteroids, one on course for Earth and one on course for Cameron's fictional Pandora. What will happen in each case?

What *could* happen in the case of an asteroid headed for Earth (if humans were not so foolish as to have fewer people than the staff of an average McDonald's searching for them—but that's another story) is that the asteroid would be detected years ahead of the collision, and the full resources of our technological civilisation could be put to work to deflect the asteroid or devise some other defence. On Pandora, life would go on oblivious of danger until the day when a brilliant light appeared in the sky and, shortly afterwards, life ended for all the people, all the animals, all the plants, all the invisible life forces Cameron imagines to exist on Pandora.

Unfortunately we humans are unable to properly assess extremely unlikely, but incredibly damaging events, such as a killer asteroid[134]. We are told one wiped out all the dinosaurs (except birds), but hey, we say, *what's the chance* one would come along *now?* We spend huge sums, usually billions of dollars, on other dangers such as road safety and aeroplane safety, while that less-than-a-McDonald's group of people plod on looking for that killer asteroid. And we have this incredible disparity of response to these two dangers because we all know just how very unlikely a killer asteroid strike would be. But in fact the probability that someone will die from a killer asteroid is in the same ballpark as the probability that they will die in a plane accident.[135] This is because, although killer asteroids hit the Earth only

---

134 The February 2013 asteroid strike that damaged the Russian city of Chelyabinsk and injured nearly 1,500 people was, of course, a 'baby', but it should remind us that asteroids are not something that we can count on to "not happen to *us*".

135 Assuming 6 billion people are killed by an asteroid every 60 million years, it averages 100 per year. Plane deaths are typically around 1,000 per year. But unlike a plane or car crash, the long-term effects of a killer asteroid last for thousands or millions of years, which magnifies their harm immeasurably.

every 60 million years or so, when they do hit, they take out almost everything: almost every living creature, plant or animal, and wipe out most of the species. But good luck to you convincing people to take precautions against asteroid strikes!

Another world-wrecking killer is a supervolcano. We alive now, who have only ever seen ordinary old little volcanoes like Krakatoa, Etna, Mt St Helens, Vesuvius, etc., find it hard to imagine the raw power, the super-massive ejection of gasses into the atmosphere, of a supervolcano. Speaking of the possibility of an eruption of a super-volcano under the Yellowstone National Park, Stephen Self of the United Kingdom's Open University said:

> "An area the size of North America can be devastated, and pro-nounced deterioration of global climate would be expected for a few years following the eruption. They could result in the devastation of world agriculture, severe disruption of food supplies, and mass star-vation. These effects could be sufficiently severe to threaten the fab-ric of civilization."

Such an eruption would cover half the United States in volcanic ash a metre deep, the emissions would block out the Sun and plunge the world into a 'perpetual winter' leading to a mass extinction. [136] At our present state of knowledge, we have no defences against a super-volcano, except to use our technology to reduce the harm if one actu-ally happens.

Other killer threats include a nearby star going supernova. A su-pernova is a cataclysmic explosion of a star, and for a few weeks the star shines brighter than an entire galaxy of thousands of millions of stars. Again, we have at present little in the way of practical de-fences. Luckily all nearby stars have been checked and none are be-lieved to be likely candidates for many thousands of years at least. This is ironic in view of the work of Svensmark on supernovas and their associated cosmic rays, which promote life in the seas. It seems we want them in the neighbourhood, but not next door. These are the kinds of reasons some speculate that planets like Earth, with long-lived, developed life systems, are very rare indeed.

---

[136] Referenced on http://www.livescience.com/environment/
050308_super_volcano.html

Some of these dangers we can or soon will be able to do something about; others we just have to use technology to cope with if they occur. But in all cases, if we as humans defend our planet or help our planet cope with any of these dangers, we will be helping our plant and animal friends as well. The planet, sooner or later, will need our technology. I know it isn't popular to see meanings in things, but often I wonder if perhaps we came into existence specifically to provide these tools to all the living beings sharing the Earth with us?

Other Human-caused Influences

Perhaps there are dangers we need to be aware of, maybe natural, maybe anthropogenic, that are not directly concerned with world temperature? For sure there are, and even though the whole subject is, like global warming, over-hyped, damage to ecosystems is very real.

Tree Clearing

Loss of natural forest cover changes living conditions for plants and animals; it alters the planet's ability to oxygenate the atmosphere; and rain patterns are altered (usually for the worse). Also, much land that was previously forest-covered is relatively poor in soil nutrients. This is so, for example, for Australia's rainforest regions. Once cleared for timber, they typically produce a few crops before becoming unproductive.[137]

The iconic loss of ice cover from Mt Kilimanjaro is a result, not of global warming, but of drying climate due to rain use changes in the surrounding areas. Most of the snowcap loss had already

---

[137] This is not to say that tree-clearing is always bad. In Australia before white settlement, the eucalypt forests were 'open'—you could walk through them. But since the disastrous introduction of lantana, this prickly shrub has closed off millions of square kilometres of land from use by native wildlife. Clearing for agriculture, or simply as the easiest way to mow down lantana, can actually make life easier for many native species.

happened before 1950, which gives global warming a complete exoneration.

For those still worried about global warming (which, as we have seen, is almost entirely a good thing), tree clearing results in land that has different reflective characteristics. Human settlements typically absorb more sunlight and are hotter than natural vegetation.

Sadly this issue is not helped by the faux-environmental lobby which, instead of sticking to facts, over-sells the damage. In Australia, this has resulted in draconian laws which not only prohibit the previously irresponsible land clearing practices, but also prevent, for example, home owners in bushfire-prone regions from making effective fire breaks around their homes. The terrible loss of life in the 2007 Victorian bushfires was a predictable result. The green movement definitely has blood on its hands.

Nitrogen from Over-fertilisation

Over-use of artificial fertilisers in order to force unnatural returns from farming lands results in wash-off of nitrogen-rich compounds into lakes and oceans. This can result in algal blooms that choke fish, and can also deplete water oxygen levels and cause death zones for fish and other water-dwelling animals.

Subdivision of Natural Wildlife Areas

By creation of networks of roads and railways and the intrusion of wind farms with their vast networks of power lines for relatively little power return, we are cutting down the range of most wildlife. The fliers (birds, flying foxes, insectivore bats, insects, etc.) have some means to resist this encroachment, but they now have to run the gauntlet of the wind turbine killing machines, so in some way or another, we are seriously harming the ability of wildlife to survive.

Unfortunately, even here, where the environmental movement has a genuine issue to which the world needs to be alerted, it is framed in inaccurate ways that could well result in poor solutions to a real problem. One key inaccuracy is the question of species extinction.

Wikipedia, for example, claim that up to 140,000 species go extinct each year. Such claims do not accord with the facts. The truth is probably nearer to one species per year going extinct due to human activity.

I am claiming a negative here (that there are not thousands of species going extinct each year), and it is impossible to prove a negative. So, to anyone who thinks I am wrong, I challenge you to pick any single year you like and actually name just *one* percent of the species that went extinct in that year. In other words, for your choice of year, tell us the names, known and recorded prior to that year, of 1,400 species that were there the year before but not the year after.

When asked to substantiate claims such as this one, those making them are forced to admit that these figures do not come from actual observations, but are calculated from unsubstantiated theoretical hypotheses (that is, models again!). The idea is that, although cats and dogs and rats and bats and almost all the various species we have discovered and named are still here, nevertheless vast numbers of species we never discovered have been disappearing. Very well, but with 140,000 species disappearing year on year, surely at least 1% of the losers should be species we have met and named? Somewhere, some time, a known and named species should be wiped out?

On occasion it has happened: passenger pigeons, dodos, thylacines, and many others—at a rate of about one per year. For one known species to disappear each year whilst 140,000 unknown species disappear, there would, statistically, have to be roughly 140,000 times more unknown and undiscovered species than known and discovered ones. But that is simply not the case. For fish, reptiles, birds, and mammals, we know nearly all the species that exist. Even for insects we know more than half of them. How could it be that extinctions just happen to happen to those species that have the *least* to do with human beings?

The reason I am labouring this point is that I believe there is a very real wrong being done here to our fellow living beings, but by making up a fake scare story, the environmental movement has missed the real and serious harm humans are doing to other creatures.

The real and tragic wrong is the actual harm inflicted upon real living, feeling individuals by human activity. Farmers erect barbed wire fences when other, safer fencing would do just as well, and bats get their wings caught in them and die a slow, agonising death. Wind farms explode the lungs of bats, blowing them up from the inside, even when they are not directly hit by the blades rotating at 300 km per hour. Baits and traps for predators to protect domestic animals result in horrific painful deaths to millions of animals.

Near my home, lizards lie on the roads to warm up (no global warming problem for them, apparently—and I live in a country that is supposedly dying from overheating) and are squashed or, worse, half-squashed, by passing cars. In a thousand ways, real sentient beings, as capable of suffering or joy as you or I, have their lives blighted by human activity. But as long as we focus our efforts on vague, impersonal (indeed fake) issues like "species loss" we never notice or correct the real and immediate damage we are doing and the suffering being inflicted upon billions of animal lives.

I don't now if you have noticed, but almost all the words we have to refer to harm to animals are bulk words, words that refer to animals as quantity or as category, almost never as actual real living individuals. That means the words are really euphemisms for something that would horrify us if put in realistic terms: "road kill", "extinctions", "species loss", "endangered species", "collateral damage", "wildlife", "native fauna", "environmental impact", "biodiversity", and so on. Where are the real, feeling, sentient, living animals in these foggy descriptions?

The use of euphemisms allows us to set an emotional distance between ourselves and the wild creatures sharing our planet with us. That in turn allows us to set their real interests aside and thrust our own emotional needs for atonement to the fore (regardless of whether satisfying our needs corresponds with the welfare of real living animals). A better world does not start with grand plans to fix "the environment"; it starts with us fixing the concrete mistakes we make that harm the real living sentient individuals all around us.

A Myriad Other Dangers

We don't have time or space here to go through dozens of other real and major ways in which humans are causing problems for our planet. And, of course, we are also causing problems for ourselves in the bargain. But we are making two really serious mistakes that will cripple our ability to actually fix any real problems, and I have explained these two mistakes above:

1. We are spending or planning to spend trillions of dollars and bankrupt our western economies fighting the entirely illusory problem of global warming; and

2. We are over-hyping and over-inflating other issues, which means that we never know what is true and what is false in anything we read. If the claim is 140,000 extinctions a year when the truth is just one per year, then as soon as people 'wise up', they will ignore *all* the claims about human damage to the environment, *including the true ones!*

Maybe There's Just Too Many People?

We seem to have entered a time of widespread self-hate. The 'meme' that humans are an evil plague upon the planet has become almost universal. Elsewhere in this book I have tried to explain why I do not find that view convincing. However, there might be a more sober case for the proposition that our current numbers are far too high, and unlike many who disagree with the climate alarmists, I have a lot of sympathy for that view. But that doesn't mean that mass starvation and suffering to get rid of the 'plague of people' is the right way to deal with our problem; in fact it could be exactly the wrong idea.

The experience of Europe and other first-world countries has been that, if people get a higher living standard and security and become confident that their children will survive to adulthood, the reproduction rate drops. Maybe a major education campaign is also needed, but at least we have proof that population control is possible when people have a decent living standard. Most of Europe is currently below the replacement rate.

What, then, would be the sensible way to reduce the human population? Can we do worse than by giving all people basic security (of their person and of necessities of life) and a decent standard of living, and educating children to understand and master the things that make the best societies work well, and thereby give parents the surety that they do not need ten children in order that two might survive to adulthood? Yes, it would be slow, but those who seek a faster way need to be very sure they know what they are asking for.

If the mere existence of so many people is bad for the planet, we should be humane to ourselves, just as we should be to animals, in trying to fix the problem.

A Sober Assessment

We have seen that carbon dioxide is plant food; it is the basis of life itself, and the more of it there is, the better plants grow and the more humans and wildlife our planet can support. We have seen that a moderately warmer planet, whilst not 100% beneficial, would certainly be more friendly to life in general and human life in particular than the planet of today; past generations feared cold and welcomed warmth; past ages thought of the warm times as the good times.

We have seen that CO_2 contributes little or nothing to global warming, and that perhaps the CO_2 concentration isn't actually any or much higher than it was in pre-industrial times. We have seen that the planet came within a whisker, during the last ice age, of not having enough of it to power photosynthesis in plants, which would have resulted in the extinction of almost all plant life on Earth, followed shortly afterwards by the extinction of almost every animal, including us.

We have seen that there is an ice age cycle which has operated for over two million years, and in that cycle the next event on the calendar is a descent into another ice age. We have seen that the ice ages are getting colder each time, and that carbon dioxide concentration follows temperature, so likely the next ice age will be colder and

CO_2 concentration will be lower and our planet will move even closer to the catastrophe point where photosynthesis stops and life on Earth as we know it comes to an end. Our industrial emissions of CO_2, far from being a problem, either have no effect or are actively beneficial in dealing with these real dangers.

Collectively our society has gone insane, agonising over a non-problem whilst ignoring real ones. This is not the way to care for ourselves and our planet, and we had better awake from our mass delusion soon, for real emergencies might come upon us at any time, and we may need our collective wits to deal with them effectively.

8

The Precautionary Principle

Okay, so far we've seen that:

- carbon dioxide is plant food, a critical necessity to life on Earth, and the more of it, the more food for humans and wildlife;
- whenever it has happened in history, warming has been good for people and life in general, and cold has been bad;
- but due to arbitrary manipulation of the data, poor temperature station siting, excessive secrecy in the climate centres, and other reasons, we don't even know for sure if all or any part of the claimed warming of the past 100 years has been real;
- the theory that any significant warming has been due to carbon dioxide fails its main experimental tests, and is therefore disproven;
- and even if CO_2 has caused the warming, the CO_2 started going up long before significant amounts were emitted by modern industry—meaning it is most likely a natural phenomenon.

We have also taken a look at the various reasons why such an obviously false idea is being pushed so hard by so many, without needing to invoke 'tinfoil hat' global conspiracy theories.

But Shouldn't We Take Precautions Just in Case? After all, the damage, if the theory just happens to be right, would be catastrophic.

This is the much vaunted "precautionary principle". Unfortunately, although it sounds good, even precautions have consequences. Let's say you're worried about storm damage to your house: you take out an insurance policy: that's common sense. But what if the policy cost you as much as the whole house? Would anyone willingly pay more in insurance than they would ever lose if the worst happened? Overly expensive "insurance" against public risks like climate change makes just as little sense if misguided 'precautions' that are worse than the original problem have damaging consequences for real people—usually not the ones making the decision. And the suffering usually falls upon the poorest and least powerful.

Let us look at just one past case in which tragic choices were taken, choices with much in common with current global warming policies.

The Cautionary Tale of Malaria and DDT

Everybody knows DDT is a highly dangerous chemical that destroys birds' eggs and kills wildlife, right?

Apparently not. D Rutledge, producer and director of *3 Billion and Counting; the death toll is mounting*, tells a revealing account of his investigations:

"Our research showed that the US, Canada, Australia, and Europe eradicated Malaria and many other insect born diseases from 1945-70 with the use of DDT. Prevention was the main or first line of defense followed by other measures — simple and easy. People were positive thinking, with an attitude of openness to space exploration, human longevity and fluidity of movement physical, mental, and spiritual.

"Something happened in America in the sixties and it was not just Viet-nam. The evidence of it is all there in the media during the Johnson, Nixon and Carter years. Suddenly, talk of research and science geared toward human betterment, anti-aging and other betterment solutions for humanity began to drop off the radar screen. Rachel Carson wrote a "novel" not a scientific study subject to peer review, but a novel which was embraced by the American public with the docility and innocence of a Jew walking into Hitler's ovens. Fear spread across the face of this land — I could see this as I read articles, journals, editorial comments, ads in the New York Times suggesting mothers were poisoning their nursing children with DDT tainted breast milk. ... People became frightened, scared, marching, regimented and mob-like.

"The mothers and fathers of my generation really were duped and still have no idea how much pain and suffering they inflicted on themselves and billions world-wide. Whipped into a frenzy of fear by Rachel Carson's Silent Spring, fearing cancer, fearing a silent spring, they were in effect, FEARING THE UNKNOWN. The people became afraid of the FUTURE, the unknown. After Earth Day in 1970, it was clear to politicians that the public was demanding getting rid of the most innocuous of chemicals — DDT. It was legally banned in 1972 by people who knew better at the time — people who were "supposed to know". People who said they were protecting us and wildlife. Were they? Did they?

Rutledge put together a team to travel the world and find out why the ban on DDT came about, and why no major news outlet has ever highlighted the terrible death toll that resulted from the ban. His team interviewed 'ordinary folk' around the world, politicians, doctors, clinics, hospitals, and many others. He continues:

"Malaria alone infects more than a half billion mostly women and children yearly with more and more deaths piled upon the deaths of the preceding years. Something was terribly wrong. This ban was touted by EPA and environmental groups everywhere as a "success" story, and is still being done to this very hour. But they are going to have to come forward with more than words, more than feel good, look good sound bites. The result of the DDT ban has been an un-speakable death toll — a sad, deception riddled tale told by people who repeat themselves and just turn their mouths on and leave while spouting a hollow litany. This sort of Tall Tale, the high story of such political stench should never be the result of a ban presumed to be life saving. Our mothers and fathers should be ashamed, then out-raged. The ones I have spoken with are.

"It took us what has seemed like forever to wade through junk sci-ence held up to be valid, to get to the hard peer-reviewed data. There have been literally thousands of studies of every possible aspect of DDT. What peer-reviewed replicated scientific data supports the ban on DDT? NONE.

"Instead we found that DDT, the most effective chemical for pre-venting Malaria and a veritable host of other diseases (West Nile virus, Lymes Disease, Rocky Mountain spotted fever, Dengue Fever, Lice, Yellow Fever, River Blindness, Elephantiasis, St. Louis Encephali-tis Virus, Typhus, Chagas Disease, Bubonic Plague, Japanese En-cephalitis, bed bugs, leprosy[138], and many others — not to mention the many bird and animals diseases which are huge in number) was a scapegoat, a glorified whipping boy that had to go down. DDT had to go down to satisfy politics, with a small "p" of course, resulting in massive reductions of [mainly] black, brown, and yellow human

---

138   The writer might to be mistaken about leprosy; some research papers suggest it increases susceptibility to the disease.

faces. ..."[139]

Rutledge, appalled by his discoveries, produced and directed *3 Billion and Counting* to document what he sees as "the greatest human death toll in the known history of man".

But still the insanity goes on. Claims that a useful chemical is dangerous go unchecked and result in action that condemns millions to death and billions to the misery of a life-long infection. But claims in the opposite direction, that humanity has discovered an effective way to combat a debilitating disease, are met with scepticism, ridicule, and even abuse. The tawdry accusation that the person defending the chemical is a paid-for shill of big business won't be far behind, regardless of the truth or otherwise of the accusation.

But is it really true that DDT has been given a bad rap? Dr. J. Gordon Edwards, Professor of entomology at San Jose State University in California, made some notes on claims made in Silent Spring:

> "[Rachel] Carson writes: "Like the robin, another American bird seems to be on the verge of extinction. This is the national symbol, the eagle."
>
> "In that very same year, 1962, the leading ornithologist in North America also mentioned the status of the robin. That authority was Roger Tory Peterson, who asked in his Life magazine Nature library book, The Birds, "What is North America's number one bird?" He then pointed out that it was the robin! The Audubon Christmas Bird Count in 1941 (before DDT) was 19,616 robins (only 8.41 seen per observer)... . Compare that with the 1960 count of 928,639 robins (or 104.01 per observer). The total was 12 times more robins seen per observer after all those years of DDT and other "modern pesticide" usage. Carson had to avoid all references to such surveys or her thesis would have been disproved by the evidence. ...

Edwards documents other problems he found in the book, amongst which was calling the hatching rate of quail and pheasant eggs exposed to DDT "few" when the research reported that for unexposed quail the result was 83.9% (not exposed) vs 75 – 80% (exposed), and for pheasants the result was 57.4% (not exposed) vs 80.6% (exposed); he writes "Perhaps she thought that her readers

---

[139] D Rutledge, Director of *3 Billion and Counting; the death toll is mounting...* (3Billionandcounting.com) at http://seekerblog.com/archives/20051012/silent-spring-is-still-killing-due-to-ddt-ban/

would never see the rather obscure journal in which [these] ... results were published..." He writes:

> "Carson's claim, therefore, that those three kinds of birds are less and less able to produce young is remarkably false—and insulting to the reader."[140]

After publication of Rachel Carson's book, Silent Spring, an increasingly active campaign was mounted against DDT, largely on allegations of harm to birds, allegations which have long been known to be groundless. And when DDT was removed from the front line of the anti-malaria fight, deaths from the disease skyrocketed. In 1970 the National Academy of Sciences (NAS) stated:

> "To only a few chemicals does man owe as great a debt as to DDT. It has contributed to the great increase in agricultural productivity, while sparing countless humanity from a host of diseases, most notably, perhaps, scrub typhus and malaria. Indeed, it is estimated that, in little more than two decades, DDT has prevented 500 million deaths due to malaria that would otherwise have been inevitable. Abandonment of this valuable insecticide should be undertaken only at such time and in such places as it is evident that the prospective gain to humanity exceeds the consequent losses. At this writing, all available substitutes for DDT are both more expensive per crop-year and decidedly more hazardous to those who manufacture and utilize them in crop treatment or for other, more general purposes."[141]

Lord Monckton claims a figure of ten million deaths per decade (mostly children) for four decades, and has been called a "loony peer" for making such outrageous statements. But as we have just seen, according to the National Academy of Sciences, the truth is far, far worse! Whatever the exact figure, a terrible amount of human misery has been caused by the needless spread of this debilitating and deadly disease. The DDT ban gives us a glimpse of the horrible cost of giving in to "feel-good" causes without taking the trouble to investigate

---

[140] Dr. J. Gordon Edwards, Professor of entomology at San Jose State University in California: The Lies of Rachel Carson.

http://www.21stcenturysciencetech.com/articles/summ02/Carson.html

[141] *The Life Sciences; Recent Progress and Application to Human Affairs,*

The World of Biological Research, Requirements for the Future. Page 432.

National Academy of Sciences (NAS), 1970.

Found online at http://www.nap.edu/catalog.php?record_id=9575.

them properly. But just being aware isn't enough: we must follow through by standing up for the truth, not for whatever popular fad one's friends, colleagues, and the media are mindlessly promoting.

Friends with whom I discuss these issues meet such claims as these with disbelief. How can it be, they ask, that so much contrary evidence exists and yet the claim is accepted so readily? It is so mind-boggling that it seems mentally easier to dismiss the evidence and accept the claim anyway.

One of the world's master liars of all time explained the psychology behind this process:

"... in the big lie there is always a certain force of credibility; because the broad masses of a nation are always more easily corrupted in the deeper strata of their emotional nature than consciously or voluntarily; and thus in the primitive simplicity of their minds they more readily fall victims to the big lie than the small lie, since they themselves often tell small lies in little matters but would be ashamed to resort to large-scale falsehoods. It would never come into their heads to fabricate colossal untruths, and they would not believe that others could have the impudence to distort the truth so infamously.

"Even though the facts which prove this to be so may be brought clearly to their minds, they will still doubt and waver and will continue to think that there may be some other explanation. For the grossly impudent lie always leaves traces behind it, even after it has been nailed down, a fact which is known to all expert liars in this world and to all who conspire together in the art of lying."[142]

This is, of course, from Hitler, and typically, he accused the Jews of it, but as a master analysis of the means by which big liars fool good people, it stands unequalled. Given the fact that believing the evidence also means accepting that one has been hugely lied to about life-and-death matters in a way that can only be called evil, most people, unwilling to impute such monstrosity to others, cave in and reject the evidence. There's little point bemoaning this fact, it is the way we humans are; the question is, what can we do about it?

One answer is that, if on some issue we are amongst the few who 'see through' the lie (as increasingly many people are doing with the global warming hoax), we must be kind to those still in the grips of

---

[142] Adolf Hitler , Mein Kampf, vol. I, ch. X.

Carbon is Life

the shysters—because, who knows, on some other issue we ourselves might still be deluded, and we would appreciate kindness and help from others in getting ourselves free, so let us do so for others when we are lucky enough to be amongst the clear-sighted.

But let us get back on track. The point of this digression into the DDT scare was to demonstrate that the precautionary principle—in the DDT case, avoiding DDT because of fears about its effects—can be worse than the problem it is taking precautions against. Such is the case with global warming, where it is seldom denied (but unfortunately seldom advertised) that the most minuscule fraction of the proposed expenditure on uselessly fighting "climate change" would be better spent, and could actually fix, any negative consequences of a warming climate. To return to the house insurance example, we noted that we would not take out a policy if the cost were as large as the value of the entire house; in the case of global warming, the more appropriate analogy would be if the cost of insurance were the value of the house, but all that the policy covered was the glass in the windows!

And, of course, should warming actually happen (and, contrary to all evidence and logic, be due to human-emitted CO_2) the benefits of warmth, which vastly exceed the deficits, will be there anyway.

Faulty Precautions in Climate Politics

Not only are the proposed precautions unimaginably more expensive than the danger warrants, but moreover they do not in fact work. *All* the precautions currently being taken in climate-related government policies are faulty, for one or another of these reasons:

- Climate policies and renewable energy policies are mixed up and conflated as if the two were the same thing, but in fact there is no inherent reason why policies designed to serve one end should automatically serve the other;
- Public policy labours under the false belief that CO_2, the basis of all life, is a toxin;
- It is accepted without question that global warming is bad,

when it is, on the whole, beneficial;

- The costs and benefits (even if one 'buys' the idea that CO_2 is pollution or that warming is bad) are not realistically assessed; symbolic acts (such as delaying global warming by a few days in a hundred years) are taken to be worth any expense (such as devastating a nation's or the world's economy);
- Things that could be fixed here and now for little expense do not even make it onto the public discussion radar; worldwide disease and famine could be ended at much less cost and with much greater benefit (even using the faulty figures from warming alarmists) than could modifying the planet's climate.

Wind Turbines: Wildlife Killing Machines

The damage done by wind turbines upon our bird and bat wildlife is one of the most under-appreciated scandals of the 'green' movement. A report by Jim Wiegand on cfact.org[143] shows in detail that a realistic estimate of animal deaths from wind farms, in the United States alone, is *39 million deaths per year*!

In the name of renewability, or of 'reducing our carbon footprint', unimaginable harm and cruelty is being inflicted upon our flying animal friends. Birds and plant-pollinating flying foxes are a vital part of the natural world. Yet for many readers, this may well be the first time you have heard that these machines are relentlessly efficient wildlife killers. Why don't we see the danger when we gaze upon a line of wind turbines?

The problem is that modern wind turbines are huge. We can only look at them and see them 'whole' from a considerable distance. As I show in a blog post on peacelegacy.org[144], this causes us to misjudge size, speed, and the forces involved. If you have ever looked down upon the ground from an aeroplane, you may have noticed this ef-

---

[143] Jim Wiegand: *Wind turbines kill up to 39 million birds a year! Big Wind hides evidence of turbine bird kills – and gets rewarded. Here's how they do it.* March 18, 2013. www.cfact.org/2013/03/18/wind-turbines-kill-up-to-39-million-birds-a-year.

[144] peacelegacy.org/articles/wind-farms-do-they-kill-birds

fect: the cars and people on the ground below look like 'toys' scooting around: we know in our brains that they are big and powerful, but they look tiny and light. And in particular, from a distance they look slow-moving.

With a huge wind turbine the problem works out like this: a turbine rotates about once every three seconds—nice and slow, right? Wrong. The turbine is so huge that this means the end of the blade is moving at around 300 kph, or about 186 miles per hour. And flying animals are fooled just as we are—they see this seemingly slow, lazy-moving blade and assume they can easily fly between two blades. But in reality they have no hope at all—a blade hits them at 300kph and slices them in half, or slices of their head, or, worse, slices off a wing and the animal spends hours slowly bleeding to death. Bats and flying foxes have it even worse; the blades don't even have to hit them to kill them because the sudden change in air pressure from a nearby blade makes their lungs explode!

Just as in the case of DDT and the ongoing disregard for human suffering, we now see the same lack of compassion towards suffering animals. Wiegand's article shows us in detail how this terrible death toll has been kept secret from us for so long: reporting rules that exclude wind farms from reporting animal deaths in conditions where a coal power station would be compelled to make a report, combined with secrecy and restricted access, and only looking for carcasses in a restricted zone much smaller than the true 'fallout' zone where carcases are likely to hit ground.

About this terrible death toll, Wiegand writes in the article cited above:

"This is intolerable, and unsustainable. It is leading to the inevitable extinction of many species, at least in many habitats, and perhaps in the entire Lower 48 States.

"Meanwhile, assorted "experts" continue to insist that the greatest threats to golden eagles are other factors like hikers getting too close to their nests, even when most abandoned nests in Southern California are nowhere near any hiking trails and wind turbines continue to slaughter eagles.

"It is essential that people realize that no energy source comes anywhere close to killing as many raptors as wind energy does. No other energy companies are allowed to pick up bodies of rare and protected species from around their production sites on a day-to-day basis, year-in and year-out. No other energy producer has a several thousand mile mortality foot print (the highly endangered whooping cranes' migratory corridor) like what wind energy has."

Not many hundreds of years ago, a completely misplaced fear of witches resulted in thousands of innocent deaths. For some decades and still ongoing, fear of "chemicals" has killed or blighted the lives of billions of people and fear of plant food is causing unimaginable suffering to animals. Is there really much difference between these cases? Have we learned anything at all about real spiritual wisdom, and genuine compassion, in all this time?

Consequences

The global warming scare is only the latest manifestation of a long history of self-destructive behaviour by our wonderful yet somehow tragically twisted species. The campaign against DDT is a solemn warning against writing off our species' tendencies for self-delusion as merely a cute or comic foible. The facts were known for most of the past forty years of the DDT scare. But did concerned activists breathe a sigh of relief that the scare was unfounded? Mostly not: the facts were ignored or even covered up, much as is now happening with global warming. And horrifyingly, some even *welcomed* the deaths of deprived children! Junkscience.com reports:

> 'Population control advocates blamed DDT for increasing third world population. In the 1960s, World Health Organization authorities be-lieved there was no alternative to the overpopulation problem but to assure that up to 40 percent of the children in poor nations would die of malaria. As an official of the Agency for International Develop-ment stated, "Rather dead than alive and riotously reproducing."'[145]

This is the kind of icy-hearted outlook on life that results when we allow our ideologies to capture us and deaden our sensitivity to the suffering of others. I define *ideologies* to be *thought patterns that simplify the world until it can all be explained by a single cause or al-*

---

145 www.junkscience.com/ddtfaq.html, referencing Desowitz, RS. 1992. Malaria Capers, W.W. Norton & Company

Carbon is Life

legedly fixed by a single solution (such as, to take a popular one, a socialist utopia).

Often thought to be harmless or even good, in reality ideologies are a defence mechanism against the desolation of our inner feelings of exile discussed in Chapter 6. As such they are a manifestation of psychic sickness, and this is why hundreds of millions can be murdered by a regime (as in the case of Stalin's Russia or Mao's China) or killed by bad policy (as in the case of the DDT ban) and yet normally good, well-meaning people can systematically turn a blind eye to the carnage and go right on feeling good about supporting those murderous people or policies.

These are deep problems all humans have to deal with in some form or another. We must never imagine that we alone (or we and those who agree with us) are immune and the problem is someone else's. All political factions, all religions, all groups of humans, have done both wonderful and terrible things at different times. Yes, some do more good or more evil than others, but we should not imagine that there is a quick fix to our problems by merely identifying the 'right' side and supporting it uncritically.

Which Precautions Make Sense?

To start with, precautions should be taken about the most likely dangers first. Expending vast public monies against a danger that is uncertain (both in the sense of whether it will happen at all and in whether it is dangerous even if it does happen) whilst actual deadly dangers such as poverty and disease are happening here and now and can be fixed at a fraction of the cost, makes no sense, either in logic or compassion.

A modern instance of the damage from violating this rule is the biofuel scandal. Indur M. Goklany writes:

"A basic contention of developing countries (DCs) and various UN bureaucracies and multilateral groups during the course of International negotiations on climate change is that industrialized countries (ICs) have a historical responsibility for global warming. This contention underlies much of the justification for insisting not only that industrialized countries reduce their greenhouse gas emissions even as developing countries are given a bye on emission reductions, but that they also subsidize clean energy development and adaptation in developing countries. ...

"But the fundamental premise behind this notion of historical responsibility is that the full consequences of fossil fuel based economic development — synonymous with industrialization — are negative. But is this premise valid?

"In fact, by virtually any objective measure of human well-being — e.g., life expectancy; infant, child and maternal mortality; prevalence of hunger and malnutrition; child labor; job opportunities for women; educational attainment; income — humanity is far better off today that it was before the start of industrialization.

"That human well-being has advanced with economic development is clearly true for industrialized countries. ... for the U.S., a surrogate for industrialized countries, ... life expectancy — perhaps the single most important indicator of human well-being — and GDP per capita — the best single measure for material well-being — increased through the 20th century, even as CO_2 emissions, population, and material, metals, and organic chemical use increased. ...

"Indeed, human well-being has also advanced for developing countries. Consider, for example, that:

"The proportion of the developing world's population living in absolute poverty (i.e., living on less than $1.25 per day in 2005 dollars), was halved from 52 percent to 26 percent between 1981 and 2005. Ironically, higher food prices, partly because of the diversion of crops to biofuels in response to climate change policies, helped push 130-155 million people into absolute poverty in 2008. This is equivalent to 2.5–3.0% of the developing world's population.

"The proportion of the developing world's population suffering from chronic hunger had declined from around 30-35 percent in 1969-1971 to 16 percent in 2003-2005. It has since increased to 18% —thanks, once again, in part to climate change policies designed to displace fossil fuels with biofuels ... The UN Food and Agricultural Organization estimates that such policies helped increase the number of people in the developing world suffering from chronic hunger by 75 million in 2007 compared to the 2003-2005 period."[146]

These figures are scandalous, and it is amazing there is not more outrage over it. In case anyone thinks this is just one person's opin-

---

[146] Indur M. Goklany. http://wattsupwiththat.com/2009/10/12/ linking-health-wealth-and-well-being-with-the-use-of-energy

ion, here is Barbara Stocking, Chief Executive of Oxfam[147]:

"Big business is piling up the profits in the biofuels bonanza whilst millions of the poorest people are suffering increased hunger and poverty as a result..."

Once again we see the pattern of forcing a policy through the political machinery until it becomes law and we can't avoid it[148], and soon afterwards (or even worse, beforehand), it is found that the policy has an horrific impact on the poor and powerless. And yet, although the people pushing these policies wear their concern for the 'underprivileged' as a badge of pride, they remain silent (or even cover up) the terrible new facts about their brainchild. Why?

One answer is a phenomenon now being called *noble cause corruption*. Writing in the context of policing, Steve Rothlein writes:[149]

"Traditional corruption is defined as the use of one's official position for personal gain. The personal gain can be economic or otherwise, such as sexual favors. As a profession, we have long understood this type of abuse of power and, when discovered and investigated, those involved are arrested.

"A less obvious but perhaps even more threatening type of misconduct in law enforcement is Noble Cause Corruption. This type of misconduct involves not necessarily the rotten apples ... but sometimes involves the best officers ..., or the golden apples. Noble Cause Corruption is a mindset or sub-culture which fosters a belief that the ends justify the means."

This idea is valuable in understanding how some extremely harmful social policies have been enacted, enforced, and have continued to hold moral high ground, despite being known for sure to be damaging. Once a 'higher good' has been identified, then any amount of damage in its name can be reconciled in the mind of a believer with practical actions of almost limitless evil.

The 'higher good' for many is the creation of a world of equality for all. From that kernel, dismantlement of the capitalist economic-industrial system can be justified (even though it, and it alone,

---

147 http://www.oxfam.org.uk/media-centre/press-releases/2012/09/biofuels-driving-up-prices

148 New South Wales, for one example, and many other jurisdictions, has a policy of mandating ethanol use.

149 http://www.patc.com/weeklyarticles/print/noble-cause-corruption.pdf

has lifted billions out of poverty). Once it is then accepted that the capitalist system has to go, its industrial base must go along with it. From there its its enabling technologies—such as effective pesticides —in turn become tainted, and replacing the 'artificial' fossil fuel by a 'natural' biofuel becomes almost a moral imperative, regardless of the misery and starvation caused by turning food into fuel.

It is sometimes objected that there are biological sources that cannot be used for food but which can be turned into biofuel. Very good. But is a legal mandate the right way to locate and use such resources? Once again, the free market provides the answer because, if costs reflected true scarcity and difficulty in obtaining a product, only those resources with no other value (such as being useful as food!) would be cost-effective as biofuel.

To deal with one objection that is sure to be raised, allowing the free market to take care of the question does not mean we have to tolerate invasion of wildernesses, etc., merely because it might be cheaper to use resources from those areas. A free market does not require that the market be unregulated. For example, a tax (or an outright prohibition) could be placed on materials obtained from wilderness areas, and then the free market would automatically price products from those areas out of the shops.

How Do We Take (Really) Effective Precautions?

The following is a small list of things we might do to be sure we are taking precautions that suit the dangers.

1. First of all, we should not be 'stampeded' into a decision because the goal is noble. A noble goal isn't enough if we harm others along the way.
2. We should *not* support policies or laws for symbolic reasons. ("Sending a message", "Doing our bit", "Standing up for the environment", etc.)
3. Nor should we judge questions of fact on any basis other than evidence. When one side of an ideological divide (political, religious, or whatever) insists that it is *immoral* to be-

lieve a certain proposition about *factual* matters (such as whether humans are causing global warming), all our red flags should go up. This is almost an invariable indicator of bad and harmful beliefs that will cause damaging decisions.

4. It is a terrible mistake to support censorship of one side of a debate. If that side is obviously wrong, then all that is needed is to give the proof of that fact. (Of course this idea is usually floated precisely because the 'obviously wrong' side is—yes—right, and their nobly corrupt opponents think themselves morally justified in hiding this fact for some 'greater good'.)

5. We can think of the precautionary principle as an insurance policy. Only take out a policy that costs a lot less than the damage we are insuring against.

6. Most importantly, we must never put ideas (and ideologies) ahead of real people (or animals, for that matter). Our planet needs its human wisdom, but those in charge need to be responsive to the concrete effects of policies instead of being wedded to the ideas behind them. A policy with fine motivations might not work in practice, and if not, we need clear sight and freedom from emotional attachment in order to modify or reverse the bad policy.

Now it is time to look at the evidence for the unthinkable—that there are many reasons why more carbon dioxide is actually beneficial.

9

Benefits of Carbon Dioxide

We have now seen that the campaign against CO_2 is entirely mistaken—and, if anything, that is a gross understatement of the terrible truth. And as for many of the ringleaders of the campaign who, knowing the truth, wilfully vilify one of the two most important molecules for life (H_2O and CO_2), it is not too strong to call it the greatest, costliest, and deadliest scam ever perpetrated in the history of the world. So let's take a short survey of the benefits of CO_2 which will be reduced or lost if humanity were ever to succeed in 're-ducing its carbon footprint'.

Global Warming—Yes, a Benefit!

This is on the whole a benefit, as we have seen. Our planet is, geologically speaking, suffering a severe chill, which has almost frozen life out from vast areas: Antarctica, Greenland, northern Canada and Alaska, northern Siberia. A few hardy species such as penguins and polar bears are making a living in those places, but the vast bulk of living creatures—including ourselves—are not able to make a decent living space in these frozen wildernesses. More CO_2 won't do much warming (notwithstanding the alarmists' claims), but even if it did, we would benefit from it, as these frigid, mostly dead area would be the ones to warm the most. And we should not lose sight of the fact that this is an observed fact of history, as we know people prospered in the Roman and Medieval warm periods and suffered misery and death in the cold Dark Ages and the Little Ice Age—unlike the unsubstantiated warmist predictions, which are entirely taken from computer modelling.

Has the alarmist millionaire Al Gore ever wondered, as his enterprises make him a fortune 'defending' us from global warming, what happened to the tens of thousands of species that once inhabited lush forests and prairies on the Antarctic continent? All gone: all that life extinguished. For each one of tens of thousands of species, some lonely creature or creatures watched as their last egg failed to hatch or their last baby died of cold before they themselves succumbed to the encroaching ice. Death, extinction, loss; that is the history of our planet's current ice infection. What insanity makes so many otherwise intelligent people cling to the present climate as if it were perfect, afraid to allow our climate to change as it always has; and if now we have a chance to bring back warmth and life, what kind of irrational thinking makes people fear it?

Resilience to Global Cooling

We have seen how carbon dioxide concentration follows temperature, not the reverse, and how, during the previous ice age, as temperatures plummeted, the CO_2 concentration fell to within a whisker of the cut-off point below which most plants are unable to grow and would, if those conditions continued for just a few hundreds or thousands of years, all go extinct. We very nearly had the greatest mass extinction in the planet's history, far greater than even the Permian extinction in which 95% of all species were obliterated.

As each ice age in the current cycle has been colder than the one before, the next one would probably (if it were not for the lucky appearance of humans and their CO_2-producing technologies) take the CO_2 concentration even lower, closer to or even below the catastrophe point. Human-emitted CO_2 has almost certainly saved the planet from this catastrophe at some point in the next several thousand years. It is a sobering thought that, although this is a simple statement of fact, it is so painful for some people that they will never be able to accept it, so strong is the need to believe in human evil.

Enhanced Plant Growth

More carbon dioxide stimulates the growth of almost all plants. That includes grasses, food crops, trees, you name it. In an era when millions of human beings have uncertain food supplies and when the biofuel fad is driving up food prices and throwing millions into food poverty, we should be happy indeed that the extra CO_2 put into the air from human industrial activity has grown enough extra food to feed several hundred million people. It has also enhanced all the plant foods for all the world's wildlife as well, giving animals a much-needed buffer against a lot of the damaging activities humans get up to.

The truth about CO_2 as plant food has been known a long time. Commercial greenhouse operators deliberately pump CO_2 into their greenhouses for a good reason: money. They make more of it selling the bigger, better plants that grow faster and with less water in a high-CO_2 environment.

The plain facts visible to anyone with a greenhouse are, however, sometimes dismissed as mere 'anecdotal evidence'. So it's a good job that thousands upon thousands of carefully conducted scientific experiments say the same thing. A very good website, co2science.org, has catalogued these experiments for easy perusal. For a few examples:

- Natural communities or ecosystems experiments show an average 48% increase in growth with a 300ppm increase in CO_2;[150]
- tree roots grow better;[151]
- more extensive roots systems allow them to withstand droughts.[152]

Here are just a sample of overview facts from co2science.org; the details are easily obtained from that site[153]:

---

150 http://www.co2science.org/data/plant_growth/dry/ecosystems.php

151 http://www.co2science.org/articles/V13/N28/B1.php

152 http://www.co2science.org/articles/V1/N2/B3.php

153 http://www.co2science.org/data/plant_growth/dry/ecosystems.php

Plant Dry Weight (Biomass) Responses to Atmospheric CO_2 Enrichment

Air enhanced CO_2 by: --------->   300ppm

Plant name	# of studies	Mean	Std. error
10 Species of Tropical Forest Tree Seedlings	1	-5.00%	0.00%
12 Species Ecosystem	3	22.70%	5.50%
12 Species From Fertile Permanent Grassland	15	67.50%	13.20%
7 Species Ecosystem	1	9.00%	0.00%
Calcareous Grassland (C3)	11	27.40%	2.80%
Calcareous Grassland dominated by Bromus erectus (Hudson)	3	28.00%	2.90%
California Annual Grassland	2	20.50%	2.50%
California Grassland of Mostly Small Annual Species	3	30.70%	18.50%
Chaparral ecosystem dominated by shrubs	1	185.00%	0.00%
Duke Forest dominated by Loblolly Pine and other trees, shrubs and vines	2	51.00%	6.40%
European Beech and Norway Spruce Ecosystem	6	45.50%	12.50%
Filamentous algae dominated by Zygnema species, but also containing some Mougeotia and Spirogyra			
Grassland Community	2	19.00%	2.80%
Grassland [High-species-diversity temperate grazed in New Zealand]	2	96.00%	47.40%
Grassland [Species-poor on a peaty gley soil]	1	50.00%	0.00%
Grassland [Species-rich on a brown earth soil over limestone]	1	56.00%	0.00%
Irish Neutral Grassland Community	4	28.50%	6.80%
Linum usitatissimum L. [Common Flax] in mixed stands with Silene cretica	2	63.50%	15.90%
Loblolly pine seedlings plus a variety of C3 and C4 weeds	2	0.00%	15.60%
Longleaf pine savannahs throughout the southeastern United States	2	50.00%	6.40%
Mixed Community of White Clover and Buffalo Grass	6	23.50%	5.20%
Mixed stand of Quaking Aspen and Paper Birch	7	76.30%	13.60%
Mixed stand of Quaking Aspen and Sugar Maple	3	122.30%	42.80%
Narrowleaf plantain/tall fescue mixture	2	50.00%	0.00%

Native Riparian Community of the United Kingdom	2	23.00%	2.10%
Native Tallgrass Prairie	3	62.00%	17.40%
Native Tallgrass Prairie Dominated by Andropogon gerardii Vitman	2	13.50%	9.50%
Natural Ecosystem Composed Primarily (more than 95% of its biomass) of Yellow Birch and White Birch	2	39.00%	12.70%
Non-grazed Grassland Near Giessen, Germany	5	23.60%	8.40%
Non-legume Dicot Component of a New Zealand Dry Sandy Pasture	1	338.00%	0.00%
Nutrient-Poor Semi-Natural Grassland	1	20.00%	0.00%
Orchard Grass and Red Clover Pasture in Switzerland	12	29.10%	5.70%
Pasture Ecosystem	1	12.00%	0.00%
Peat communities from Wales	6	104.70%	13.00%
Perennial Grassland of Cedar Creek Natural History Area in central Minnesota, USA	3	24.30%	4.90%
Perennial Ryegrass and White Clover Mixture			
Pristine Tallgrass Prairie	3	23.30%	12.20%
Scrub Oak Palmetto Ecosystem	3	69.30%	20.80%
Semi-Natural Grassland	3	27.30%	11.40%
Semi-natural grassland in central Sweden	4	34.30%	12.20%
Semi-Natural Grassland in Minnesota, USA	1	17.00%	0.00%
Shortgrass steppe in NE Colorado	13	34.50%	4.20%
Silene cretica in mixed stands with Common Flax	2	98.50%	17.30%
Southern China Ecosystem Of Six Tree Species [Tree ecosystem]	4	28.80%	4.60%
Tree ecosystem [Southern China Ecosystem Of Six Tree Species]	4	28.80%	4.60%
Understory Plants in a Spruce Model Ecosystem	3	33.70%	13.40%
Understory Vegetation of a Sweetgum Plantation	1	3.00%	0.00%
Wetland Communities Comprised of Native Graminoids and Reed Canary Grass	4	41.50%	10.00%
Overall Average:		48.84%	

These figures show almost a 50% increase in plant dry weight when CO_2 is increased by 300ppm. Like it or not, the world's population is still increasing and will likely increase by more than a billion before it stabilises. How can anyone carelessly throw away a benefit of the order of 50% more plant growth—with all the associated increase in food production—merely because computer models that are known to be wrong predict global warming of a magnitude that has happened in the past and was demonstrably beneficial when it did happen?

The world is greening. Human industrial CO_2 emissions are the reason millions have been saved from starvation. The great physicist Freeman Dyson, who has sometimes been lauded as Einstein's successor, said:

> "It's certainly true that carbon dioxide is good for vegetation, About 15 percent of agricultural yields are due to CO2 we put in the atmosphere. From that point of view, it's a real plus to burn coal and oil."[154]

In this light the motives, and indeed the sanity of those pushing the alarm must be seriously questioned. For example, in the face of such stupendously good news as we have seen above, the Chicago Field Museum Climate Exhibit put up a display for kids entitled: *CO₂ makes Poison Ivy grow*[155]. Of course it does! Poison ivy is a plant just like all the world's food crops, all benefiting from enhanced CO_2. For another example, the journal *Nature* blames growth of marmots (yes, seriously!) on global warming[156]. Be they animals, plants, bigger, smaller, fatter, skinnier, more of them, fewer of them, you name it, any change in this huge and varied world of ours is labelled bad and blamed on global warming and CO_2. Truly, the inmates are in charge of the lunatic asylum.

Plants Cope Better with Environmental Stress

Just as we might expect if plants are growing bigger with more CO_2,

---

[154] Quoted in *Climatologists are no Einsteins, says his successor*, by Paul Mulshine. See http://blog.nj.com/njv_paul_mulshine/2013/04/climatologists_are_no-_einstein.html

[155] http://wattsupwiththat.com/2010/07/18/why-ill-never-take-my-kids-to-the-chicago-field-museum/

[156] http://www.nature.com/news/2010/100721/full/news.2010.366.html

they also show better resilience to all kinds of environmental stress. Such stresses include:

- high and low temperatures,
- pollution,
- disease,
- drought.

As a consequence, all kinds of good changes are happening. The ranges of many of the world's forests are increasing—despite the damaging acts of human beings. One major study showed 25% in total net primary production for a 200ppm increase in CO_2.[157] This is an actual experimental result, not a computer model.

Also it appears that plants' medicinal properties are improving. Two researchers write: "Elevated CO_2 has also been demonstrated to increase the biomass of plants grown for medicinal purposes while simultaneously increasing the concentrations of the disease-fighting substances produced within them. It is likely, therefore, that the on-going rise in the air's CO_2 content will continue to increase food production around the world, while maintaining the nutritive quality of that food and enhancing the production of certain disease-inhibiting plant compounds."[158] We have now seen a range of evidence that the world's food security is being increased by improved plant growth. If only the biofuel hysteria had not stolen that benefit from the poor by putting their food into wealthy nations' automobiles!

The Planet is Turning Green!

All this isn't just theoretical, and it isn't just conclusions drawn from experiments in a laboratory. Our planet is turning green and it has been actually measured.

In a peer-reviewed study, Ramakrishna R. Nemani and colleagues

---

[157] DeLucia *et al*, Net Primary Production of a Forest Ecosystem with Experimental CO2 Enrichment. http://www.sciencemag.org/content/284/5417/1177.short

[158] Idso & Idso, Effects of atmospheric CO2 enrichment on plant constituents related to animal and human health.
http://www.sciencedirect.com/science/article/pii/S0098847200000915

write:

> "Recent climatic changes have enhanced plant growth in northern mid-latitudes and high latitudes. ... We present a global investigation of vegetation responses to climatic changes by analyzing 18 years (1982 to 1999) of both climatic data and satellite observations of vegetation activity. Our results indicate that global changes in climate have eased several critical climatic constraints to plant growth, such that net primary production increased 6% (3.4 petagrams of carbon over 18 years) globally. The largest increase was in tropical ecosystems. Amazon rain forests accounted for 42% of the global increase in net primary production, owing mainly to decreased cloud cover and the resulting increase in solar radiation."[159]

That 6% represents enough extra food, in a population of 6 billion, to feed 360 million people. That's a measured increase over only 18 years. (In passing we might also note how much credence we should give to scare stories about the Amazon forests being decimated by global warming.) This period also represents the greatest warming episode in our time, as 1998 was the year with the peak temperature since the Little Ice Age.

We have a choice: we can believe the evidence of actual measured plant growth in the real world, or we can believe scare stories based on computer models that have never made a successful prediction. The question we need to ask ourselves is: how much credibility should a claim of future danger have, and how massive a future danger should it be, to convince us that we should surrender actual existing real-life food for 360 million people? That's the amount of real harm we will do to real struggling people in the poorest nations if we succeed in undoing 'climate change'.

How Will We Feed 9 Billion People?

That's the question this inevitably leads us to: nine billion human souls on Earth by 2050—how will we feed them? This whole issue is often presented in the media and by alarmists as if the climate realists

---

[159] *Science* 6 June 2003: Vol. 300 no. 5625 pp. 1560-1563. DOI: 10.1126/science.1082750
http://www.sciencemag.org/content/300/5625/1560.abstract

Carbon is Life

are ignoring the precautionary principle discussed in the previous chapter by ignoring the heating effects of CO_2. But as we have seen, the proper use of this principle isn't to take the worst prediction possible and spend unlimited amounts defending against it, regardless of cost, because those costs have their own consequences. Britain's biggest lobby group for the elderly, Age UK, claims that 400 pensioners each week freeze to death during winter, simply because they cannot afford to heat their homes[160]. Expect this number to go through the roof when the real costs of disfiguring Britain's magnificent natural areas with unreliable and super-expensive wind farms hits the poor in future years.

The misuse of precaution is having deadly effects right now, not only in poor nations but in the affluent United Kingdom too. These people's lives are not "eggs that must be broken" to make some utopian green omelette. The need to care for each and every innocent life is an indispensible prerequisite for a decent and caring society, and I discuss it in the final chapter.

But back to feeding nine billion people. Tragically, all the "carbon abatement" efforts and carbon taxes and other schemes being rammed through in Australia and elsewhere will make our chances of doing that much worse. The Center for the Study of Carbon Dioxide and Global Change has released a study that, unlike most alarmist studies, takes the extra plant growth due to CO_2 into account. It points out "Global food production must increase by 70 to 100 percent by the year 2050, if we are to adequately feed a global population of nine billion people at that time." It continues: "... at the International Conference on Rising Atmospheric Carbon Dioxide and Plant Productivity, it was concluded that a doubling of the air's CO_2 concentration would likely lead to a 50% increase in photosynthesis in C3 plants, a doubling of water use efficiency in both C3 and C4 plants, significant increases in biological nitrogen fixation in almost all biological systems, and an increase in the ability of plants to adapt to a variety of environmental stresses..."[161]

---

160 http://british-news-portal.co.uk/freeze-to-death-peril-for-the-elderly
161 Estimates of Global Food Production in the Year 2050: Will We Produce Enough

The report attempts to separate out the two major causes of increased food production: improved technology and other processes, versus yield-enhancing effects of greater atmospheric CO_2. It concludes that from the first factor on its own, only Europe is predicted to be food-secure. But from the additional food growth due to CO_2, North and South America also will become *possibly* (their emphasis) food-secure.

That paints a dismal picture of our future prospects for feeding ourselves within forty years, and we might not buy in to the assumptions on which the report is based. After all, we have had two hundred years of failed predictions of resource depletion disaster. But regardless, the point is made crystal clear that whatever difficulties we face in future, we will be much, much better placed to grapple with them if we have a CO_2-enriched atmosphere to help us.

Affluent westerners patting themselves on the back for 'reducing their carbon footprint' really need to face these hard questions. And even if, after considering the flimsiness of the case for dangerous 'climate change' and the rock-solid case for immense benefits from CO_2, western nations still decide to push ahead with 'carbon abatement'— we must still ask if we have the right to do so at such immense cost to others far worse off than ourselves?

Carbon dioxide, along with water, is one of the most critical components of the web of life on Earth. Calling CO_2 'pollution' is insanity—or worse. Declaring war against carbon is equivalent to declaring war against all life on our planet.

to Adequately Feed the World? Craig D. Idso, Ph.D. Center for the Study of Carbon Dioxide and Global Change. 15 June 2011. http://www.co2science.org/education/reports/foodsecurity/GlobalFoodProductionEstimates2050.pdf

Carbon is Life

10

Saving Ourselves, Wildlife, and the Planet

So far we have seen how a scare campaign has been invented about a trace gas in the atmosphere which, far from being 'pollution', is actually one of the most critical, essential components of all living things. From the most primitive microbe to a human being, the two biggest components of their living tissues (around 94%) is carbon dioxide and water—food and drink for plants. The energy from the Sun converts these two chemicals by photosynthesis into the living tissues of plants, and then when animals feed they convert plant material into their own. Indirectly, we are made of carbon dioxide and water, plus a few other essentials like calcium and trace minerals.

We have seen how life on land thrives in the warmth, and how today's planetary temperatures are nowhere near as high as they have been, even in historical times such as the prosperous times at the height of the Roman Empire. We have seen how cold times cause death, starvation, disease, misery, wars, and species extinction. Our generation, the first and only in the history of the world to fear warmth, would be called insane—and rightly so—by any of our ancestors. Our descendants will have far worse than that to say about us, for they will have to live with the terrible results of social policies devised at the greatest cost to fight—not merely a non-existent problem, but an anti-problem—in other words, to fight against something good.

So what should we do? How should we get ourselves out of the mess, and even better, get really sound environmental policies in place?

Don't Confuse Renewable Energy with Climate

How often have you heard it said "Save energy—you'll be doing your bit for climate change as well"? I certainly have heard it countless times. An advertising campaign right now on Queensland television promotes electricity efficiency based on exactly this confusion between saving energy (and therefore saving non-renewable resources such as coal) and fighting the scary but non-existent climate change monster.

This confusion is more damaging than we might think. Those of us who have acknowledged the life-giving properties of CO_2 will know that "reducing your carbon footprint" and building "green" power stations are actually bad for the planet because they reduce CO_2 emissions, thereby restricting plant growth and hence restricting the food supply for humans and wildlife.

But even those of us who are wise to the global warming hoax must still face the consequences of using up non-renewable energy supplies. There are some opponents of the hoax who deny limits to the planet's capacities, but we should not be amongst them. An exponential growth of human population must eventually exceed the planet's limits. Those who say otherwise can and do point to a long line of failed predictions of disaster, starting well before Thomas Malthus in the late eighteenth century and proceeding through the Club of Rome in the 1960s and on until we reach today's eco-environmentalists. But the fact that these predictions failed does not mean that the Earth is infinite, nor that humans can expand their numbers without end.

Confusing these two issues messes up our thinking whether we have yet seen through the hoax or not. Those who still believe in the hoax will applaud "green" coal on renewability grounds, insensible to the fact that powering the sequestration of carbon dioxide (even if one thinks that is a good thing) takes a lot of energy and uses up even more non-renewable coal.

Once we separate these issues (regardless of our opinions about each one separately) we can ask other important questions. For example: for whose benefit are we pursuing policies:

a   concerning climate change?

b   concerning renewable energy?

The answer, I believe, is different for the two issues. Supporters of the green movement (and, indeed, perhaps a majority of people in western nations) support carbon dioxide emission limits because of concern for the world at large: biodiversity, species loss, harm to living being—including, but not limited to, humans.

On the other hand, no such motive can exist for pursuing renewable energy. Not another species on Earth would care, if they could understand the question, whether, deep in the Earth, reserves of coal and oil have been left intact or mined out. They are useless assets for every species except *homo sapiens*. This is strange because many of the more extreme supporters of renewability actually hate humanity (we have seen examples earlier); there is even a society for the voluntary extinction of human beings[162]. A television 'documentary' was even made[163] showing approvingly how the program makers envisaged Earth would 'recover' if humans were to suddenly vanish. Paradoxically, most of these same people are zealous supporters of recycling programs, etc., which have no use for anyone except human beings. Such is the confusion coming from mixing up two different problems. To summarise:

- Renewability is a human-centric problem; we do it for our own benefit, not for 'the planet';
  - a realistic understanding of the role of carbon dioxide leads to efficient use of non-renewable resources, but
  - belief in the 'climate change' scare leads to crazy ideas like "clean coal" that will waste even more resources solely in order to power the damaging burial of valuable CO_2 plant food.
- "Climate change" is an issue in the minds of most people for altruistic reasons, such as the welfare of the planet as a whole;
  - but even so, is it not better to understand the real risks

---

162 Voluntary Human Extinction Movement. http://vhemt.org.

163 *Life After People.*

(such as too little CO_2 to power plant growth, and the possibility of a new ice age) and the real benefits (such as more food, a greener world, and the fact that warmth is on the whole better for life than cold) so our decision-making is as good as it can be?

Reject All Schemes to Restrict Carbon Emissions

We have to get firm with our politicians about this. In Australia, a federal election gave the "Greens" a share of power. A carbon tax has now been imposed. The government is 100% in favour of the eco-religion and its hatred of the carbon basis of life on Earth. The leader of the opposition, not believing in the CO_2 scare, persuaded his party to stop supporting the proposed emissions trading scheme. But he nonetheless pays his obeisances to the climate god by telling the Australian voters that his non-policy as just as good at 'reducing our carbon footprint' as the government's carbon tax would be.

We need to stop being willing slaves to the rituals of the eco-religion, stop using its terminology, jargon words like: "carbon footprint", "carbon pollution", "emissions" (as if sulphur dioxide and carbon dioxide were equivalently nasty) and so on. Bad ideologies must be tackled head-on, not opposed half-apologetically whilst allowing their proponents to choose the framework for the debate.

We need to be so opposed to schemes to reduce carbon dioxide emissions that we actively *increase* our own donation of plant food to the atmosphere whenever that doesn't increase usage of non-renewables. Burn our own paper rubbish, or persuade our local authorities to burn it rather than bury it (preferably using the waste heat to make power, because we *all* agree that saving non-renewable resources is a good thing for humanity). And we need to be proud of it, and actively defend these actions against all who thoughtlessly criticise them.

How we do that needs wisdom, of course. It might involve denouncing an arrogant politician who has just labelled you a 'denier' or 'in the pay of Big Oil', but more likely it will involve recognising the genuine altruism of most of those who embrace the green movement. They became 'green' because they wanted to help the planet, help wildlife, help the poor, conserve the planet's resources. Those are all noble goals, and it is a crime—but not theirs!—that their good will has been subverted to what is ultimately, on the part of the ringleaders, a combination of political motives and self interest.

Stop Buying into Scare Campaigns

The global warming scare wasn't the first and it won't be the last. As its hold weakens, some new fright will be invented. This is quite deliberate, and completely cynical. We of good will should not be so naïve as to think that everyone is as well-meaning as ourselves. We have seen already that, as the warming scare slowly sinks in the west, a new scare, 'acidification' of the oceans, is being hyped, with the same victim, CO_2, wrongly framed for a crime it didn't commit.

Keeping CO_2 in the cross-hairs is deliberate because it is the perfect scapegoat. It provides ideal justification for increased laws and social control which the more political of the scare supporters desire, whilst a market that trades in *not* doing something (i.e. not emitting CO_2) is the perfect hunting ground for financial market rogues who trade, trade, endlessly trade, doing no real good in the world, and ending up with vast fortunes that were ultimately skimmed from the savings of decent people working hard for a living and actually contributing to the welfare of the world. What better 'commodity' for fraudsters could there possibly be than the *absence* of a real commodity? Financial traders earning more money in a day than most of us in a year, are amongst the most vociferous supporters of carbon trading.

Oppose the Use of Ethanol for Fuel

We have seen how burning food in cars for westerners takes that food from the mouths of the truly poor. The website C3headlines.-

com points out that[164]:

- Ethanol produces little or no additional energy versus energy needed to produce it.
- It can damage vehicle engines not designed to run on ethanol.
- It result[s] in greater CO_2 emissions than fossil fuel.
- Using food for fuel causes rising food prices either directly or by competing with food crops.
- Food riots and hunger have been [a] direct result of higher ethanol production.
- It encourages clearing of climate-stabilising forest lands.
- It increases the use of fertiliser, leading to greater runoff and NOx emissions.
- Huge amounts of scarce fresh water are wasted to produce a single gallon of ethanol.
- It produces less energy than simply burning the biomass to produce electricity.
- It is only commercially viable with government subsidies and forced-use mandates.

C3headlines' final point again highlights the distortions that are produced by use of governmental diktats instead of allowing individuals to optimise their own lives. Biofuel is a feel-good, but socially-delinquent luxury for comparatively wealthy westerners. We who are fortunate enough to live in safe, well-off western nations really need to 'get over it' and start doing the things that actually help the poor and the planet, instead of the things that do no good except to give us nice, warm, fuzzy, but undeserved feelings about our own altruism.

Scientific Research

Until recently, the scientific system for allowing the best ideas to bubble to the top has been the peer review of scientific work before publication in scientific journals. I think it is fair to say that the op-

---

164 http://www.c3headlines.com/renewable-energybiofuelsethanol

Carbon is Life

eration of this system has been smashed by the unprincipled way it has been subverted by climate 'science'. It will take a while to be recognised, but I suspect this system cannot be repaired.

Scientific peer-review has enjoyed uncritical esteem bordering on adulation for quite a while, and it is not generally understood that it is not necessarily the best way to sort good science from bad. In the first place, it is not and never was inherently foolproof; but more importantly, to the extent that it did work, it relied upon moral integrity—for example, a paper should not be rejected because its conclusions do not support the reviewer's political beliefs; the editorial boards of journals should not be 'stuffed' with believers in a political ideology pretending to be scientists.

To this writer at least, the climate 'science' community has shown it is utterly bereft of the necessary integrity to make this system work—in fact they seem to want it *not* to work. We have seen papers rejected after undue influence; a 'rebuttal' of good but sceptical work published with no right of reply by the original authors; an editor forced out of his position; scientists dismissed for speaking out. We have seen data and code hidden, data manipulated, freedom of information requests subverted. Peer review can no longer be trusted to sift the good from the bad[165].

In some ways this is probably a good thing. Science that depends upon highly complex computer calculations cannot be adequately checked for accuracy (let alone for deliberate falsification!) by the superficial check for reasonableness that academic referees typically give to papers. Only complete publication of all data and all computer programs (not merely the algorithms) suffices to ensure that such research *can* be checked—and even that doesn't guarantee that it

---

165 A recent, perhaps even more shocking, example of this is taking place even now. A well-researched paper by Spencer and Braswell cast doubt upon one of the lynchpins of the global warming theory: the idea that clouds do not cool the earth (yes, really!). The journal editor resigned because the paper was published, despite its having passed a fully competent review, and a 'rebuttal' was rushed through the supposedly independent peer review system in record time. You can follow developments in this story at http://wattsupwiththat.com/reference-pages/the-spencer-braswell-dessler-papers.

will be.

The recent experience of blogs like climateaudit.org has shown that good science can be done in an entirely different framework. I suggest that some way of rating 'open science', meaning science conducted in 'real time' with online criticism in the associated comment stream, might be the path that science takes in future. One day, perhaps, surviving a blog comment 'gauntlet' might be the measure of the worth of scientific writings? There is nothing sacrosanct about peer review. Socrates established his philosophy by talking in the Athenian forum, Newton his science by writing a book, Einstein by publishing papers. The next great scientist might announce the breakthrough on a blog.

Independence for Scientific Research

The failure of peer review is by no means the end of the failures of science during this sorry episode. It has also shown us how a vicious feedback can be set up in which some scientists proclaim a coming disaster, politicians get worried about it, and then funding primarily goes to those who continue to provide 'evidence' confirming the disaster prediction. This generates much more alarm, which generates more funding, and soon an upwards spiral of increasing fear and ever more funds to generate yet more fear is firmly established, involving government grants, enthusiastic scientific supporters, financial market supporters who think they can profitably 'feed' off the policy outcomes, a frightened but sincere public in all honesty believing they are helping the world, politicians getting re-elected with the aid of the swing voters who come their way when they demonstrate their 'green' credentials, the owners of Big Coal willing to take the insults while they take their profits to the bank, and so on.

Highly damaging social phenomena can gain a life of their own, and our future depends on fixing the broken processes which lead to it. This broken system of funding scientific research is one of them. Examining in detail how this can be corrected is not within the scope of this book, but there is always room for a few suggestions:

- Break the connection between political objectives and science and science funding.

- ○ Funding bodies should have independence from politicians.
- ○ Funding requirements should favour 'devil's advocate' research over 'more of the same' research whenever total funding on a topic exceeds a certain threshold.
- Repudiate the newly established idea that scientists should be activists. Insist that scientists never use their paid work time for activism: either stick to science or get another job. Scientists who have an axe to grind cannot do good science.

Stand Up for the Web of Life

It is difficult to explain just what I have in mind here. I am sure all environmentalists would say they stand up for the web of life; and yet, as we have seen, almost to the last one, they are supporting anti-life, anti-carbon policies which serve the ends of bankers, market traders, Big Coal, bigger government, and control freaks of all stripes. So how do we know we are really working in the interests of life, rather than selfish vested interests?

The early communist leaders called the rank and file who supported them "useful idiots". Is this the attitude of the ringleaders of the environmental movement towards their own supporters? Looking back at the terrible damage being done to life in our own time by the anti-carbon madness, and combining this with the disregard of the alarmist leaders for the practical recommendations of their own message, it is hard to give them the benefit of the doubt. When Al Gore buys an expensive apartment by the sea shore, the same sea he claims will be flooding its banks within a few years, it is hard to credit that he means his message. When the U.S. President's wife takes a lavish holiday in Spain, with all the concomitant luxury air trips, whilst supporting policies that sap even the most minor luxuries from the lives of ordinary folk, it is hard to credit her genuineness.

So we have to accept, I think, that we are infiltrated—almost every environmental organisation on Earth has indeed been taken over—by ill-purposed people wearing the mask of holiness. What can we do about it?

I have one suggestion, a new way of seeing and valuing our fellow living creatures, which may help 'inoculate' us against the plottings of the false friends who have so badly misdirected the environmental movement, a movement which should have been a force for life and for happiness for all living creatures. I leave this suggestion to a chapter of its own, following this one. For now, let me stick to some more practical points.

People are different. Jung devised an interesting personality theory, which has since been refined into what is now called the Myers-Briggs test. It shows that people differ in at least four specific personality 'dimensions'. One of them is the thinking-feeling dimension. This measures whether a person primarily uses a personal or an impersonal standard in making decisions. To take a somewhat over-simplified example, a judge who says "how can I fine this poor person who has had to put up with so much hardship?" is probably a 'feeler', whereas a judge who says "like it or not, he stole that money deliberately and must pay the penalty" is probably a thinker.

Few of us are pure 'thinkers' or pure 'feelers', though some of us come close, but most of us tend to feel more at ease one side of the middle than the other; and whichever end of this dimension one's own personality leans, one tends to misunderstand the choices of those who lean the other way. This leads to a lot of misunderstanding and even ill-will, when in fact both attitudes have their proper place. We each need the ability to see things the 'other way round'; and also society needs both kinds of people to be well-grounded.

One way this affects us is that a lot of 'thinkers' who have studied global warming quickly come to the conclusion that there is no evidence in support of the theory (and many will go further and note the abundance of evidence against it). They then go out proclaiming "Nothing to worry about! Forget all about the problem." This is quite sound as far as it goes, but it doesn't go nearly far enough to satisfy the concerns of a 'feeler'. The feeler will see the thinker as a heartless person taking risks with the future of the world because, however much we study a problem, we can never be *certain* that we understand it; and maybe that killer planetary heatwave might engulf us despite our reasonings?

It seems that we simply must try to be more understanding of those who differ from ourselves. If we are a thinker, we need to practise asking ourselves 'feeling' questions. *What if I am wrong? How about those polar bears?* And so on. Feelers need to deliberately practise taking their feelings into the thought dimension. *Okay, the thinkers may indeed be right about this all being a beat-up, and if so, perhaps there are other dangers I should be concerned about, such as the food crisis which would follow any reduction in carbon dioxide?*

This amounts to saying we have to be ready to enter our *discomfort zone*. There are lots of kudos and social approval and even jobs and promotions available to those who support the *status quo* policy of anti-life carbon hatred. If we stay with our preconceptions and our comfort-zone thinking style, will we be able to see through the deception? Does a real lover of life and nature want a feel-good pat on the back at the cost of real harm to the web of life? Let us actually *do* good, rather than merely *feel* good.

11

Amidst the Chaos:
Seeds of a Higher Civilisation

Where there is no vision, the people perish. "[166]

It seems paradoxical, but it is often true, that the times of greatest danger are also those of greatest opportunity. We, that is, humans as a species, survived the past great ice age, perhaps barely scraping through at one point some seventy thousand years ago[167]. We slowly built on our successes as the current interglacial brought life-giving warm times, passing from the stone ages to the bronze age, the iron age, and lately the industrial and the information ages. And through it all, our closeness to the other life of our planet has slowly faded. Our sense of peril from sabre-toothed cats or our fear of hunger from failing to hunt an aurochs has receded as our power over nature has effloresced.

And now we really do have the power to mess up our planet well and truly. And in that lovable, but sometimes tragic way of ours, we condemn the things we've got right and ignore lots of things we are doing poorly. We have already seen how the global warming hysteria and the DDT tragedy have 'fought' nonexistent problems at the cost of actual lives amongst the poorest and least powerful.

On the other side of the ledger we could mention a host of real

---

166 Proverbs 29:18.

167 A good account of the various population bottleneck theories can be found in Stanley H. Ambrose, *Late Pleistocene human population bottlenecks, volcanic winter, and differentiation of modern humans.* Journal of Human Evolution V 34, Issue 6, June 1998, Pages 623-651.
http://ice2.uab.cat/argo/Argo_actualitzacio/argo_butlleti/ccee/geologia/arxius/1
Ambrose%201998.pdf

problems that are not addressed or are addressed badly, such as over-use of water resources, cutting up habitats and disrupting ecosystems, damming important fish migratory streams, emitting *real* pollution such as oxides of nitrogen (NO, NO_2, and others) and of sulphur (SO_2, SO_3, and others), polluting the oceans, using weapon-strength sounds to harm or kill fish and aquatic mammals, releasing species into foreign ecosystems where they do damage, and releasing chemicals into the environment during mining and industrial operations. Then, of course, we might blow ourselves up with a nuclear weapon or two, which would do none of us any good. We sure aren't getting it right yet, but what if we did? What if we found a way to turn it all around and use our ingenuity to fix the *real* problems?

To get started on that plan, it is unlikely to be enough to merely identify the real problems (although we seem to be very poor at even completing that step). We would also have to know where to look to find *root causes* of the real problems.

What (if Anything) Lies Beneath?

It would be careless to move on without asking a key question: Does anything at all underlie all the various real problems confronting humanity? Maybe each problem is simply an isolated phenomenon? Land degradation might have causes unrelated to those of over-crowded cities and social decay, which might again be unrelated to chemical poisons in the environment.

In one sense I believe the various issues are indeed *un*related: I do not believe there is a single overarching 'explanation', a single 'solution' which would fix everything. In many ways, ideologies serve as a one-point focus that narrows and damages human thought. Whether it be a religious, a political, or even a 'scientific' ideology, the all-absorbing totalising effect of trying to fit all the multifarious doings of human beings into a single overarching explanation is, I am sure, harmful to our capacity to follow the evidence wherever it leads and to uncover the processes that operate in human doings. And this failure will often have serious consequences.

But the mistake of fitting the whole world into the picture painted by a single ideology is only one extreme: the other is to assume that everything that happens is completely *ad hoc*, unrelated, and accidental; and, as a result, to only mount a day-by-day fight against the surface appearances of our problems. The truth is most likely somewhere in between: that there are, after all, certain widespread commonalities underlying much of our strife. One example is perhaps the ease with which we can become fanatical about something or another. Could there be an underlying reason for that?

Perhaps so, but we might well be pessimistic about our chances for finding the deep causes for our human foibles. As we have seen, even 'experts' are thoroughly messing up straightforward matter-of-fact questions such as "Has the planet been warming up?" What, then, are our chances of finding good answers to intangible questions like "What, at a deep level, are we doing wrong?", let alone ambitious questions like: "What makes for a good life?" "How do we generate happiness and eliminate cruelty, poverty, and injustice?" "How do we bring ourselves into harmony with each other and with the natural world?"

Hard though these issues might be, we had all better make a start to finding answers to them. And we will have to look in places we have mostly ignored up to now, or perhaps feared to look, because the answers certainly aren't where we have been looking: in politics, science, social engineering, economics, technology, or any other concrete discipline. Those are where many of the mistakes happen, but there are always hints of something more universal 'behind' them—for example, in so many fields, personal vanity prevents those in charge from accepting new ideas and analyses. In one form or another, this seems to be a constant for as long back as we have historical records. Perhaps, then, just talking the particular issues might not locate the root causes of our mistakes.

That means we have to take seriously some uncomfortable questions about the operation of our minds and emotions, about the human spirit, even about our genetic make-up and perhaps some blind spots in our understanding of ourselves that result from that. We might even have to ask 'ultimate' questions: is there purpose in the universe? even: Is there a God?

No, don't worry, I won't even be attempting to discuss such questions in this book; but I make the point that these topics must become acceptable subjects for polite public discussion. One important reason why is that, as long as we can't raise these emotive subjects in public discussion, we fall prey to having our emotions and beliefs hijacked anyway, and being unable to talk and work together to understand what is happening to us.

For example, it is not too far from the truth that the world is now dominated by a small handful of powerful ideologies. The West, for example, is almost entirely dominated in the intellectual sense (and also to a large degree in the practical sense) by a single, but complex and multifaceted, ideology, one of whose branches is an eco-religion with a forty-year track record of creating feel-good emotions about the environment in its followers, but simultaneously allowing them to have a murderous indifference to the deadly effects upon real people and real animals of the very policies they feel so good about.

The critical point here is that this ideology is based around concepts that used to be religious, such as finding a purpose in life, and developing the 'right' sort of inner self. It used to be that we needed to be 'saved', or 'find nirvana', and become a monk or nun or send out missionaries, and so on; but now, we are required to be 'authentic', and 'culturally sensitive', to hold 'green' values and disapprove of bourgeois ideals like making a living and raising a family. My point is not that this is all bad, but rather that, unconsciously, we all are indeed raised as followers of a very detailed and prescriptive ideological system. It is not all bad—but a lot of it is—and critical discussion of it is unacceptable or worse. 'Sensitive' questions have to become available for discussion without provoking hostility, or we won't be able to talk about solutions.

It Has Happened Before (and It Wasn't Pleasant)

It may be that this generation is living in the twilight of the greatest time of openness and freedom our species has ever experienced[168]—a time inspired by the Enlightenment and by the ideals of the American Revolution, by a (genuine) scientific philosophy of advancing hypotheses and testing them against reality. The dictatorial powers of the 'old'—and now discredited—theistic religion had waned, allowing human thought and individual freedom to flourish. But recently the growing power of the new, atheistic, but just as intolerant, politico-religion is steadily removing our freedoms as it intrudes itself into all the political, intellectual, and moral spaces that have been vacated by the old religions of god and angels. Thus, after a few brief centuries of living with the right to hold our own opinions and speak our own minds, we who are alive today in Western society (meaning the great democracies of the world, including India) may well be too naïve and complacent for our own good about ideologies that may one day abolish the freedoms we hold most dear.

We would be wise to take a lesson from our own past about the long-term effects of such doctrines. Once a decline has set in, it can take centuries, or even millennia, to recover. Lester Thurow, in *The Future of Capitalism*, also raises the possibility that we are heading into a new dark age.[169] About the difficulty of getting out of one, he writes:

"In the Dark Ages as now, there was no vision of how one made a better life. They knew that standards of living had been higher in the past, but they were too disorganised to get back to the past or to organise a march to the future."[170]

He adds:

"They had, or could have had, all the technologies possessed by Rome, but they did not have the values to generate the organisational abilities that would have been necessary to recreate what had previously existed... Values, not technology, dictated that they would sit in the Dark Ages century after century."[171]

We face a corresponding lack of some critical factor today. The

---

168 This much was apparent to Karl Popper when he wrote *The Open Society and Its Enemies* during World War II.

169 Thurow, p263

170 Thurow, p 267.

171 Thurow, p262.

old certainties of religion no longer play a significant part in western public life, and in those countries where religion does take a dominant role, it is more commonly a relapse into fundamentalism and obscurantism than a force for enlightenment. In *The Disappearance of God*, Richard Friedman considers how belief in God has become increasingly irrelevant to our life as a society, and he asks:

> "On what else could we base a morality if not on God? One has to have been comatose through this [i.e. 20th] century in order not to know that this is no longer just a theoretical point, not just a really good question... It is a matter of the fate of our species, being on the brink of discovery and of destruction at the same juncture in our life."[172]

Clearly the old religions have become unbelievable for a controlling portion of society's intelligentsia. But the replacement non-theistic ideology that moved into the vacated places in our psyches and which now exerts a controlling influence amongst movers and shakers is not meeting the challenges confronting us: it is clearly taking us away from sound science, from ideals of individual freedom, from an open free society, and—far worse!—from compassion as well.

Part of the trouble is that often the things that matter most can't be measured (security, community spirit, fulfilment, quality of life, environmental preservation), and so political decisions hinge instead on only those things that can be (employment statistics, GDP, taking visible action, etc.), but which are an inadequate guarantee of well-being. And as we have seen, when environmental and other intangible issues are taken into account, it is usually done from the most anti-scientific, politically-motivated, and emotive perspective imaginable. Thus our political and social leaders are taking us headlong down a closed-minded path like the one that led from the Roman Empire to the Dark Ages.

In the 1980s it was often claimed that ethics was a luxury we couldn't afford. Now, apparently, ethics is a necessity we can't obtain at any price, as is demonstrated by the disastrous and deadly outcomes of bad ideas like demonising DDT and running an hysterical "the sky is falling" propaganda war based on climate pseudo-science,

---

[172] Friedman, p275.

a war against the technologies and social understandings that have lifted large segments of humanity out of poverty, and against the molecule that is the basis of life on Earth. Ethics is a stuffy subject, and gets maybe five enrolments in a large university; and yet our political and economic dilemmas are fundamentally a failure of ethics. As Friedman emphasises, in the past

> "...belief in a supreme power that could... offer a faithful follower a measure of security helped to make life livable. "I shall fear no evil, for you are with me"... The growth and spread of the feeling of divine absence contributed to producing an age of uncertainty, insecurity, and vulnerability. Second, much of morality was based on beliefs that "God has told you what to do." Beliefs ranged from general principles ("Love your neighbour as yourself") to very specific commandments ("Don't oppress the widow and the orphan")."

Friedman explains how the sense of a divine power guiding humanity has gradually receded, culminating at the end of the nineteenth century in Nietzsche's declaration that "God is dead." The happenings of the twentieth century have certainly given us something to ponder. We have had the two worst wars in history, the deliberate attempt to exterminate whole races, and the development of weapons that can obliterate all life on earth a hundred times over, along with a crumbling of social relationships since World War 2 and the apparent loss of the understanding of the reasons why the great democracies were places worth living—the best places ever in the history of the world for pursuing happiness and finding opportunities for personal growth. Is the problem that we cannot find a genuine vision for civilisation, and so we have accepted all manner of false ones instead?

Visionary Findings—or Finding a Vision

"Without vision the people perish." But the problem is the vast differences amongst us: some are traditional believers, some humanists, some atheists, some politically inspired, and so on. What kind of vision could possibly unite us to the point where we can collectively sustain a world civilisation in the long term?

I think it is fair to say that so far (meaning the past few hundred years) modern western society (whose ways and wisdoms I shall call "Westernism" for short) has been sustained by common beliefs which we all hold for uncommon reasons. For example, most of us believe murder is wrong. Some believe this is because God said so; some because it is unethical; some see it as a medical condition or the result of poor upbringing; some see murder as against enlightened self-interest; and so on. But enough of us agree murder is wrong that we can agree in common to have laws against murder and a police force to catch murderers, and courts and prisons etc. And yet there is no deep, unifying understanding or common wisdom held by all of us explaining why murder is wrong.

This seems to be a strength of democracies. Despite having few deep beliefs in common, we can all agree on practical action by attending to the practical part rather than worrying about the reasons others agree or disagree with us.

This is quite an old thread in the collective psyche of the West. In the time of Elizabeth I of England, the country was weak from the heresy hunts and the terrible bonfires of Elizabeth's fanatical sister Mary. Although Elizabeth had bonfires of her own, she usually chose to ignore people's inner beliefs provided they did the required things, saying she had "no desire to make windows into men's souls"[173]. Although religious prejudice took some centuries to die down, the idea that we can get along despite disagreeing on important issues has taken hold and sustained Westernism for some centuries.

Recent events, such as the ideologically motivated undermining of science discussed earlier in this book, but also including quite different developments such as religiously inspired terrorism, have now posed a critical threat to the health of Westernism, perhaps even to its continued existence. As Westernism is what gave each of us a good life and has been, as its ideals spread around the world, steadily pulling hundreds of millions out of poverty, this should deeply con-

---

[173] http://www.royal.gov.uk/HistoryoftheMonarchy/KingsandQueensofEngland/Th eTudors/ElizabethI.aspx

cern us.

But how can I advocate finding a vision for our society when I have already pointed to the destructive effects of ideologies, which corrupt by seeking the single, overarching 'explanation', or 'vision of a perfect world', or some such?

I admit it's a subtle line I am drawing, but I believe it is a valid one. *"We hold these truths to be self-evident, that all men are created equal, that they are endowed by their Creator with certain unalienable Rights, that among these are Life, Liberty and the pursuit of Happiness. —That to secure these rights, Governments are instituted among Men..."*[174] The American Declaration of Independence promulgated a vision in the sense I am using the word. It is worth looking beyond the emotional appeal of the text and spend a few minutes teasing it apart.

Firstly, the vision it promotes is not an incontrovertible truth, however self-evident it may appear. Perhaps all men are not created equal? Perhaps we are not entitled to life, or liberty, or the pursuit of happiness? These things cannot be proved, and the Document wisely does not attempt to do so; but we can be invited to participate in the vision, to agree to grant to each other a right to life, liberty, and the pursuit of happiness. But beyond that, nothing.

We are not told to accept socialism or capitalism, to be or not be Christian or Muslim or atheist, and so on. In that silence is the genius of this Document, that it gave a vision but not an ideology. The wisdom to find that fine line separating a vision that can inspire from an ideology that enslaves, and to stop at just the right point; that genius, I believe, is the critical missing factor in the world today. The leaders who strut the world stage in our time have by and large bought into some ideology or other holus-bolus, for no one wants to be found without a vision and most are not wise enough to understand what a worthy vision should be.

---

174 The Declaration of Independence of the Thirteen Colonies. July 4, 1776.

Some Suggestions

"Vision" suggests something noble, far-reaching, powerful, inspiring, enabling, life-affirming. Our shared vision should ideally be all of that, of course, but it might be that no one of us is sufficiently noble, far-seeing, powerful, inspiring, enabling, or life-affirming to come up with that vision all on our own. Just as the Declaration of Independence, although it had a primary author (Thomas Jefferson), was in the end the work of a committee, so it may be that our collective vision for a safe, life-affirming, peaceful future world might need to be a collaborative effort.

Here are some thoughts.

Suggestion I:

Firstly, we need our vision to be long-term. A hundred years? Why not a thousand? A thousand? Why not a million? A million? Why not the entire remaining life-span of the planet Earth?

Because, some might say, we won't be around in a thousand years, let alone a million, let alone hundreds of millions. True, but our descendants can take care of the challenges of their time, and change whatever might be imperfect in the vision we bequeath them, provided we leave them a sound legacy: safety and security, good institutions such as workable governments, liberty, decent education, a healthy scientific infrastructure and understanding of the genuine scientific method, good public infrastructure, a sound environment, engagement with the natural world and the inspiration of seeing our planet's beautiful wild creatures flourishing in their natural habitats, peace amongst all humanity, the end of cruelty to animals, and much more.

I am sure you have your own thoughts on what needs to be added to the list. But unless we rise out of short-term thinking and the demoralising mindset it leads to, we are unlikely to make the most of our potential. Short-term thinking leads to even shorter-term solutions. We need to have a vision that is greater than ourselves.

The next few suggestions might seem to be airy-fairy stuff—

sounds good, but not really relevant or practical. So I'll give a few examples to show why they really do matter and make practical sense.

Suggestion II:

Whatever policies we devise, *we should be ambitious enough to try to benefit everyone.* We do that already, you ask? I think not. 'Bottom line' thinking, 'the big picture' thinking, 'the national interest' thinking, these are amongst the most common thought patterns for decision-making in the world today. And they are not noticeably selfish in themselves. They conform to the idea of a *greater good* for which we all, it is argued, should strive.

But the greater good implies there might be a lesser good that is being trampled; the national interest implies the existence of a multitude of individual interests that could be sacrificed in pursuing the overall good. The Bottom Line will be mostly influenced by the biggest of the individual lines, and the least by the lines belonging to the poor, weak, or powerless. But what else can we do? Surely we have to get the best overall result we can? To many this seems the obviously correct ethical way to go about living. Indeed, it seemed as much to me until, some years ago, circumstances beyond my control brought me face-to-face with the problems inherent in it. In suggesting we should not be satisfied with partial solutions, however much they maximise overall benefit, I am making a radical suggestion.

Radical, but not original!

Jesus taught it when he said that whatever one does to the least of God's creatures, one is doing to him, and He taught it again in the parable of the shepherd[175].

Muhammad taught it when he said that killing one innocent per-

---

175 "What do you think? A man has a hundred sheep, and one of them has gone astray, does he not leave the ninety-nine on the hills and go in search of the one that went astray? And if he finds it, truly, I say to you, he rejoices over it more than over the ninety-nine that never went astray. So it is not the will of my father who is in heaven that one of these little ones should perish." (Matt 18:10-14)

son is a bad as killing the whole world[176].

Socrates taught it when he refused to cooperate with arresting for execution the innocent Leon of Salmis, knowing that Leon would be arrested anyway and the only effect of his refusal would be to add his own name to the list of victims[177]. "Evil may be done," one might say, "*but not by me.*"

Gandhi taught it with a lifetime demonstration of non-violence.

Buddha taught it with his message of non-violence and practical ways to eliminate suffering[178].

To cut a long story short, we have been getting a continuous message from the best and wisest humans who have ever lived, telling us to stop looking at the bottom line and start being concerned for the welfare of each and every sentient being. We should not be satisfied with partial solutions; we should not 'weigh' one person's or one group's interests against those of others. We should try for the stars.

But the obvious question is: will we succeed? And the obvious answer is: maybe sometimes, but for sure not all the time. But of this we can be certain: if we try to benefit everyone without exception, then if or when we fail, we can at least say honestly to those whom we failed: "We did our best; we did not deliberately choose to sacrifice your interests for the rest of us." That is the one thing that just so very many 'rebels' against society needed to hear to soothe the pain of the various wrongs that had been done to them, but we never told them because we couldn't.

I have mentioned one of the advantages of democracy earlier, but this highlights one of its disadvantages. In the words of the aphorism, democracy is two wolves and a sheep voting on what to have for dinner. 'Weighing' one group's interests against another and maximising

---

[176] "...he who slayeth one, unless it be a person guilty of manslaughter, or of spreading disorders in the land, shall be as though he had slain all mankind; but he who saveth a life, shall be as though he had saved all mankind alive." (V.35. – trans. [Rodwell])

[177] Plato: Apology

[178] "A man is not a great man because he is a warrior and kills other men; but because he hurts not any living being he in truth is called a great man." (The Dhammapada. 270 – trans. [Juan Mascaro])

Carbon is Life

the overall good often ends up benefiting the wealthy, powerful, or numerous and harming the weak and 'insignificant'. Although democracies tend to account for the interests of many more people than other forms of government do, they are not infallible in caring for the few in number or the very weak. In the recent economic crisis, millions of ordinary people lost everything they had, but only banks and the tycoons who ran them were deemed "too big to be allowed to fail".

Suggestion III:

My next suggestion is related to the previous one, but despite appearances it is not the same point in reverse: *Whatever we do, we must not deliberately harm even a single innocent.*

The two ideas are related: if I am trying to harm an innocent I can't be trying to benefit everyone, and vice versa. But they are different: I can avoid trying to harm anyone without trying to benefit everyone. But despite this difference many of the teachings mentioned earlier from the wisest human beings also cover the ground here too. The two ideas are complementary parts of a whole, and that whole is in fact the moral dimension that used to be known in a more innocent age as the difference between good and evil.

How do these ideas help us?

The ruling moral idea of the modern world, the idea of maximising some kind of benefit, such as happiness over misery, or whatever (there are a multitude of fine variations on the idea) is known philosophically as utilitarianism. It is definitely better than a whole host of primitive moral ideas, such as putting one's sins upon a scapegoat, or sacrificing to a deity, or killing others for having the wrong colour or religion or nationality: the advance that utilitarianism represented in moral thinking should not be minimised. Combined with ideals of democracy, freedom of speech, personal choice, and so on, it has revolutionised the world compared with that of only a few centuries ago.

I doubt if many of us alive now would prefer to live in ages past.

There were roles: for different classes or castes, or for the sexes, or roles determined by one's parentage; these used at one time to determine the life of an individual even before birth, long before any personal tastes or skills could have become apparent. Who today wants that? Although the pendulum is of recent decades swinging back once more towards collectivist control of our lives, there are still many who understand the value of the philosophy of individualism that protested against these roles and, in effect, said "Spread the right of choice, of being able to pursue one's own idea of happiness, as widely as possible." Utilitarianism played no small part in inspiring many of the changes that made for the best features of the modern world.

But, as in past ages, humanity is now at a turning point where a greater and deeper insight is needed, where a 'bulk quantity' ethic of merely maximising some good thing is now letting too many small things fall through the cracks. The marginalised and the weak, whose collective welfare doesn't amount to a blip on the radar of the wealthy Western consciousness, are now suffering because they are invisible to Westerners concerned with their own problems.

And those Western problems need not be selfish ones, as the DDT tragedy shows us: Westerners concerned with being good environmentally-conscious planetary citizens have allowed forty million deaths, mostly of children, to happen on their watch. And now we are implementing ideologically motivated laws aimed at restricting the amount of CO_2 plant fertiliser available in the atmosphere to feed the growth of food for humans and wildlife—a public policy of dealing death to the planet, threatening, if it could succeed, a planetary famine tragedy of unimaginable magnitude some time in the near to medium future. And if our ethical insight stays in the utilitarian groove, if we don't learn a lesson from the great souls who have been loved, respected and followed by a majority of people who have ever lived, we will never have the will, as a community and as a planetary civilisation, to stop these destructive policies in their tracks—*before* they cause the death and misery!

Synergy of Goodness

Let us take a few moments to see how the three suggestions in the previous section work together. I hasten to add that some of what I am about to say I cannot prove; it is mere speculation. But inasmuch as it is what it claims to be, it is sounder than all the climate pseudo-science bunkum to which the world is now enthralled as it rushes towards disaster.

Suggestions II and III together form an ethical philosophy: try to help everyone, never try to harm any innocent for any reason.[179] This ethic is immediate: unlike utilitarianism, we do not have to compute far-off consequences of our actions to know whether they add to or subtract from the combined happiness of the entire world. That is the crippling mistake of utilitarian ethics because the calculation isn't even well-defined. (How do we compare the relative value of one person's delight against another's misery to know whether it is 'worth' making one suffer for the other?) But even if there were a clearly defined formula for happiness, who could calculate all the consequences of their actions to ever know what the upshot of them will be?

But that is just what the climate activists are trying to do: they believe it is worth depriving underdeveloped societies of the same advanced industrial technologies that enhance our own lives, and depriving our children and theirs of the same access to cheap energy that we have enjoyed, because arbitrarily parameterised computer-modelled calculations of some uncertain future state of the world say so. And sometimes they come right out and say as much in words. I won't name names, but you can easily find out for yourself, that top key movers and shakers in the world want all of these things to happen to you:

- One top Royal wants to be reborn as a virus that kills a large proportion of humanity[180];

---

179 This philosophy is called the Principle of Goodness. Papers on this philosophy can be found at http://principleofgoodness.net.

180 "If I were reincarnated I would wish to be returned to earth as a killer virus to lower human population levels."

- a top UN official hopes that the world's advanced civilisations collapse[181];
- a highly respected environmentalist who became a legend in his own lifetime hopes hundreds of thousands of people per day will be 'eliminated'[182];
- a top-selling new-age author and biologist thinks that cannibalism is a "radical but realistic solution to the problem of over-population".

And so on. Such icy-heartedness comes about, not because these people are naturally evil, but because they mistakenly think that the end justifies the means; that there are 'acceptable losses' in pursuit of a sufficiently lofty goal. That these are not merely silly things said in the heat of the moment is proved by the fact that ten million real people per decade have actually died in the wake of the DDT ban, which has been tolerated for forty years. These are not hypothetical computer projection 'victims' like the imaginary death tolls in the global warming 'scenarios'; they are real, sentient, human beings whose lives have been considered 'worth' sacrificing for the 'greater good'. We need to be able to look into the eyes of a single suffering child and say "Enough! No 'greater good' for the world can justify your misery."

The ethics of living a good life in the here and now, with compassion for every creature—every child victim of the DDT ban, every bird or bat killed by a wind turbine—combines with Suggestion I (thinking for the long term) to give us the ability to find a vision that is free from the heartlessness of ideologies. We can believe in sustainability, for example, without accepting that it requires us to leave billions in energy poverty, withholding from developing peoples the benefits of the energy-rich technologies that lifted our own lives out of grinding, demoralising life-long hard work.

---

[181] "Isn't the only hope for the planet that the industrialized civilizations collapse? Isn't it our responsibility to bring that about?"

[182] "It's terrible to have to say this. World population must be stabilized, and to do that we must eliminate 350,000 people per day. This is so horrible to contemplate that we shouldn't even say it. But the general situation in which we are involved is lamentable."

Carbon is Life

This point was brought home to me a few years ago when I visited the Pendon Museum in England. Pendon is a model railway depicting an historically accurate representation of the country and the life of rural England before World War 2. The villages and country are modelled to such a high accuracy as to show the mirror on the wall in a bedroom, or a robin resting on a spade handle in a garden. When mechanisation had delivered railways but cars were still for the very rich, when washing machines were unthought-of by 'ordinary' folk, the lives of the common people were filled with hard, grinding, exhausting drudgery for both men and women—and this was in a first-world country, not a village in Africa or India, where conditions even now can be much worse.

Moderns, especially eco-environmentalists relaxing in their armchairs surrounded by every amenity, can spend their daydreams idealising the lives of the energy-poor. They can even—knowing they can return to the modern life at any time—give up their 'corrupt lifestyle' and grow their own food and ride their bicycle instead of driving a car in their toy version of an energy-poor lifestyle. But they can never live the actual grinding life of a genuinely poor person, they can never truly experience the fear of a crop failure or a drought, because they know they can always return to 'civilisation' should they need to. And the leaders of the movement, lecturing us on our damaging lifestyles, the Al Gores and the Dukes of Edinburgh, do not themselves even play at being poor; they burn through energy at a rate that could empower the lives of thousands of the genuinely poor.

What, then, could we do? We could take the actions needed in the here and now to help others, whilst planning for the long term. There is ample non-renewable coal and oil to fuel the world for a few hundred years while we develop reliable renewable energy sources such as fusion power. We should and must encourage all societies to stop multiplying the number of human beings. Once a good life, a life of security with a reasonable guarantee that a new-born child will most likely survive into adulthood, has been achieved, the cultural habit of having more than a replacement number of children must be changed. If nothing else, we want our world to have room upon it for our non-human friends as well as ourselves.

We must believe in quality rather than quantity, aiming for a world in which every single person can live a rewarding, fulfilling life, even if, through education and security, the birth rate falls and there are fewer people on Earth in future than now.

That future world, without poverty, war a thing forgotten, will be a world where every child can find his or her own life passion and have a decent chance of actually achieving success. It will be a world that has no need of ideologies that must always find a 'villain', dividing the world into 'us' and 'them', sacrificing the few for the benefit of the many. It will be a world where humans will not hate humanity to satisfy their need to reconnect with nature, because the entire human world, inspired by a vision of a flourishing, healthy, long-lived planet, will be living permanently in connection with the living spirit of the planet.

The ethical vision will protect the long-term vision from becoming blind to real sentient beings in the here and now, whilst the long-term vision will protect the ethical vision from any short-sightedness that might otherwise cause us to choose an unsustainable path.

Our planet could and, I believe, will, become a world in which everyone can believe in the welfare of everyone else because each one will have had a childhood in which they, previously, were cared for and loved and protected. As our own safety and welfare improves, we will find it ever easier to spend some of our compassion on others rather than on ourselves.

It might take many generations to do this, because most of us alive now have already been damaged; perhaps we need our selfishness to stitch the wounds in our own psyches. But if we can find just enough extra love and extra compassion to make sure that the next generation can live lives even a little bit more secure and more inspired by love than our own lives have been, then we can set the world on an upward spiral. Our children will be able to do better than we have, to find better ways to help themselves and help others, and to find better ways to bring human life and the natural life of our planet into ever greater harmony.

From that point onwards, the sky will be the limit. But we have to ensure that our children are not prevented by the dead hand of the past from taking that very first step on the upward spiral. To us falls the job of eliminating all the 'isms': all the preconceived notions that categorise people and blame them simply for being who they are and who they cannot avoid being. The 'isms' that demonise whites, or males, or Westerners, or the rich, or the bourgeoisie or the entire human race, that blame individuals here and now because some ideology puts the faults of the past at the door of some category they belong to, all these 'isms' have had their day; they can contribute nothing to a happy, flourishing life-affirming future planet.

According to the ancient scriptures, mankind was expelled from the Garden of Eden; innocence was lost and our bond and our harmony with the planet that created us was torn. Torn but not broken. The loss of innocence and our separation from other creatures was the start of knowledge. It is the knowledge that permitted us to wipe out whole ecosystems and develop nuclear weapons, but it is also the knowledge that will permit us to protect the planet from rogue asteroids and other such dangers requiring solutions beyond the capabilities of any other species on Earth. We are, and our planet needs us to be, the thinking, tool-making, language-using, record-keeping, technological, star-gazing, story-telling, civilisation-building species that nature made us.

Our path is forwards, and one thing we will need, perhaps more than any other, is an environmental ethos based, not on regret for harmony lost, not on a wish to return to ignorance and to surrender our technological heritage, but on the aspiration to discover a new harmony with all of Earth's life by fully developing our knowledge and our wisdom, and choosing our path with love.

Appendix
Resources

To Follow the Debate

I am a firm believer in the idea that each of us should 'see with our own eyes'; that is, don't take anyone's word for it, including mine. Sadly we live in a society that is riddled with misinformation, both innocent and deliberate, on all kinds of issues. In the old Soviet Union, the newspaper Pravda (Russian for truth) was known for its lies. Similarly, with issues like DDT, global warming, and others, we have found, tragically, that even the hard core scientific enterprise has already fallen to political manipulation and expediency.

It cannot have escaped your notice that I think that the main pushers of the global warming alarm are not merely mistaken, but are outright lying. Unfortunately, when one group is bare-faced lying to you about something, a very good cover is to accuse the ones telling you the truth of lying. To defend themselves, the truth-tellers will then tell you the liars are lying, and in short, you are confronted with two groups each hurling accusations against each other. Such is the case in the climate debate; the uninitiated bystander can't trust anyone without at least checking up on some of the claims at issue. It would be naïve to simply trust whoever happens to occupy the seats of power at the time—or, if you prefer to be a rebel, those who do *not* occupy them!

I was lucky—I have a scientific background that helped me to single out the key questions and to quickly see that the catastrophic anthropogenic global warming theory (CAGW), aka 'climate change', aka 'global weirding', aka 'climate disruption', aka heaven-knows-what-it-will-be-called-tomorrow, was disproved by the facts. If you, too, know some basic physics, you should be able to do likewise. Even if not, I hope that in the preceding pages I have laid out the major issues well enough that you will have an idea what key points you want to check for yourself.

The resources I shall list here come from the climate realist side of the climate debate; the reason for that is simple: the realists (sometimes called skeptics) almost always allow uncensored comments on their articles from anybody at all; so anyone can post comments disagreeing with their evidence, their reasoning, or their conclusions. The major alarmist sites, on the other hand, are virtually all strongly censored against criticism of AGW. If you can't read the criticisms, then you can't be sure there isn't some fatal flaw in the argument. Even if one didn't have the wealth of evidence against the AGW theory, some of which I have included in this book, the fact that the major AGW promoters can't tolerate criticism should tell you something.

Web Sites

wattsupwiththat.com

The essential climate realist site right now for making up your own mind about the whole global warming question has to be Anthony Watts' *wattsupwiththat.com*, usually abbreviated to WUWT. WUWT publishes everything from jokes to scientific articles. It doesn't try to hide the other side's arguments, and even publishes some alarmist articles and press releases and simply lets them loose for the site's army of commenters to tell you what's amiss with them.

co2science.org

co2science.org is a site that collects and collates scientific papers concerning CO_2 and its effects on living things. Corals and terrestrial plants are the subject of a great many papers, all published in accepted peer-reviewed scientific journals. If anyone has been claiming there is "not a single" peer-reviewed paper demonstrating the benefits of CO_2, then this would be a good site to check whether you have been correctly informed.

climateaudit.org

climateaudit.org is the site owned by Steve McIntyre who, with Ross McKitrick, used their expert statistical knowledge to demolish the hockey stick graph, which, ignoring the medieval warming and the Little Ice Age, purported to show that temperatures were basically flat for centuries until humans started putting increased CO_2 into the atmosphere. Climate Audit is usually a more technical site than Wattsupwiththat, although it also publishes some non-technical articles.

plantsneedco2.org

Collected information on the necessity for CO_2 for life. I hope that one of the major messages readers will take from this book is the positive importance of CO_2 to life. It isn't just a case of "No need to worry, no need for that carbon tax." No, it is a case of "How dare you tax the food of life itself?!"

surfacestations.org

This site was organised by Anthony Watts, from wattsupwiththat.com, to conduct a survey of United Stated temperature measurement stations. The shocking results—stations in the path of aircraft exhaust, or surrounded by a car park, or beside an air conditioner exhaust—made for embarrassing reading by the official US temperature networks.

joannenova.com.au

This is Joanne Nova's very entertaining website. It has a cutting edge bite to it if you like your reading spiced a bit. And you'll find Joanne's "Skeptic's Handbook" in which she tells skeptics how to defeat alarmists in arguments. If you are still wondering whether the skeptics are a bunch of well-organised, financed-by-Big-Coal agent provocateurs, this handbook should be enlightening. (Hint: it's all about evidence or the lack of it, not about tricking people.) My favourite line: "There's a point about cost-benefit here. How many people are we willing to kill in order to protect us from the unproven threat of CO_2?"

icecap.us

This site is a compendium of news, divided into six categories (three visible at the top, three more you have to scroll (way!) down for). A bit hard to find old articles, but lots of the latest news without having to search for it.

peacelegacy.org

This is my own site about the more general problem of how we get a world at peace—permanent peace. There are quite a few articles there about global warming though—you have to put out the brushfires as well as stop the cause of the conflagration.

wingedhearts.org

My wife Gitie and I have been blessed to be taken into a family of Australian Magpies, be shown their secrets and learn some of their language. This site is all about the wonders of our non-human friends.

principleofgoodness.net

This small site, also mine, is a reference point for scholarly articles about the ethical inspiration for the thoughts I shared in the final chapter of this book.

Books

By Ian Plimer

Heaven and Earth; global warming: the missing science. Connor Court Publishing, 2009. Ian Plimer is a geologist and a courageous man who previously came to prominence for an earlier book, "Telling Lies for God", and when he sued creationists for misleading the public. He lost on a technicality. An account of that controversy can be found at http://www.noanswersingenesis.org.au/creation_science_and_free_speech2.htm. His current book is the place to go for in-depth scientific information about our planet and its temperature history. A warning though: in a massive work it is almost impossible to make no errors at all, and the few that Plimer has made have been widely misrepresented by alarmists as if to show that the entire book is incompetent; it isn't; judge it for yourself and take no one's word for it.

By Robert M. Carter

Climate: The Counter Consensus. Stacey International 2010. Robert Carter is a highly qualified geologist and climate scientist and an expert witness on the subject. This book is also solidly scientific, but perhaps more accessible for lay people than Plimer's.

By Garth W. Paltridge

The Climate Caper. Connor Court Publishing, 2009. Garth Paltridge is yet another qualified scientist—an atmospheric physicist. This book is short and non-technical.

By Ian Wishart

Air Con: the (seriously) inconvenient truth about global warming. If you want to learn a bit about the human story and plots and agendas and so on, Ian Wishart, journalist, might be to your fancy.

By Mark Lawson

A Guide to Climate Change Lunacy: bad forecasting, terrible solutions. Connor Court Publishing, 2010. Mark Lawson, a senior journalist, gives us another take on the science and the human story.

By James Delingpole

Killing the Earth to Save It. Connor Court Publishing, 2012. Also published under the title **Watermelons**. James Delingpole is a UK columnist with an entertaining style who isn't afraid to disagree with the establishment. His book is about the social cost of the bad policies. I can't resist giving you a quote:

> "This has always been the great challenge for those of us trying to explode the great AGW myth: you're grappling not only with complex facts but also with emotions as powerful and deep-seated and irrational as you'd find among the adherents of a religion.
>
> "AGW is a religion. It has its high priests and prophets: Al Gore, the Prince of Wales, Tim Flannery, Ross Garnaut. It has its temples: The National Academy of Sciences, the CSIRO; the IPCC. It has its warrior monks (and nuns): Leonardo DiCaprio, Cate Blanchett. It has the concept of original sin—the Carbon Footprint—which can be bought off with the help of indulgences—Carbon Offsets. It is motivated by an overwhelming guilt that we are all sinners but that we can redeem ourselves through mortification of the flesh (e.g. replacing bright light-bulbs that work with flickering, yellow, eco-friendly ones that give you a headache but are apparently better for the environment, so long as you forget about all the mercury they contain) and self-abnegation (taking fewer holidays, paying more money to the government, sacrificing present pleasures for the sake of future generations).
>
> "And most important of all it is based on no hard evidence whatsoever. Only on faith. Pure, blind faith." (p. 249)

By Matthew Sinclair

Let them Eat Carbon; the price of failing climate change policies, and how governments and big business profit from them. Biteback Publishing Ltd, 2011. All about the terrible social results of these planet-suicidal policies.

By Andrew Montford

The Hockey Stick Illusion. Stacey International 2010. The full and shocking story of the rewriting of the world's temperature history.

Hiding the Decline; a history of the Climategate affair. Createspace 2012. If you thought the hockey stick itself was bad enough with its bad statistics and the shenanigans surrounding the publication of supporting papers, the Climategate affair, which I have largely skipped over in this book, has made things a whole lot worse. The famed proxies that were the basis of the hockey stick do, in fact, turn around and decline from the late 20th century onwards, but this fact was hidden from us in the plot of the hockey stick. If they don't follow the temperature now, when we can check them against thermometers, why should we think they followed the temperature in centuries past, when they can't be checked?

The Other Side

I apologise, I have nothing to recommend for you from the pro-alarmist side of the question, for the simple reason: I know of none that I consider to be both honest and competent. This for the simple reason that I believe the CAGW theory is so clearly and definitively wrong, illogical, and contrary to the evidence all around us, that intelligent, competent people continuing to advocate it either haven't done their homework—or worse. This isn't a genuine scientific debate between two hypotheses, each with something to recommend it. That much should be clear from the alarmists' long-term attempt to never let a debate get started by the simple method of declaring it to be already over.

But you must, if you are new to the whole issue, check things for yourself. To this aim, you might want to check out:

climatedebatedaily.com

This site has two columns, "Calls to action" (pro-alarmist) and "Dissenting voices" (realist views). Most stuff published on either side ends up in one or the other column.

Finally...

I wrote this book after my own investigation into the subject, having no view beforehand except the "default" one: I had unthinkingly assumed the alarmists were right because "everybody" knew it. When I learned that the official opinion was wrong and, in the case of many of its proponents, fraudulent as well, I was still in two minds about writing my own book—many other excellent works were already available or becoming available. I finally decided to do so because the key message I saw from the realist side was "No need to worry, the danger isn't so great." But that is by no means the whole story. The alarmism isn't merely asking us to do something unnecessary, it is asking (or, with taxes and regulations, *forcing*) us to do something that is highly damaging (and might even be fatal) to ourselves, to our animal friends who share this planet with us, and to this wonderful world itself. Carbon dioxide is plant food. The more of it there is, the better; humanity simply doesn't have the capacity to raise it to a dangerous level even if we burned all the fossil fuel that exists.

But when this scam, the greatest hoax ever perpetrated against a gullible public, eventually falls apart, as it must, because the climate won't cooperate in the long run, we cannot let our guard down. When some new issue is invented to justify shutting down western civilisation or human technological or social development, it likely is just the next step in the long-term attack upon human happiness, however plausible it might at first appear. The ideological mindset that created this hoax will put together another one; the goal remains the same but the excuses for it keep on changing. Whatever new "humanity-is-evil" meme is foisted onto us tomorrow, please don't take it on trust; check it out. Our planet needs us to use the wonderful brains it has blessed us with.

Book Web Sites

carbonislifebook.com
bunyagrovepress.com

Index

abstraction 132

academics 34, 129, 140

acidification 1, 30, 52, 54, 55, 193

activity, human 38, 83, 109, 110, 142, 157, 158

Africa 40, 41, 217

agriculture 32, 42, 62, 63, 154, 155

AGW 27, 35, 37, 108, 110, 222, 226

Al Qaeda 31

alarmism 28, 33, 50, 70, 84, 95, 97, 99, 106, 137, 228

alarmists 10, 17, 27, 29, 35, 37, 44, 50, 51, 57, 60, 67-69, 146, 148, 159, 170, 179, 186, 223, 225, 227

alcohols 16

alienation 134, 138-140

alkaline 52, 54, 55

Amazon 10, 46, 186

America 10, 144, 154, 164, 166, 188

American Revolution 205

Americas 43

amino acids 16

Ammonia 52

An Inconvenient Truth 36, 80

animals 7, 10, 11, 17, 19, 20, 22, 26, 55, 64, 114, 123, 127, 132-134, 136, 138, 141, 147, 153, 155, 156, 158, 160, 165, 171, 172, 177, 181, 184, 189, 204, 210

wild 10, 138

Anthropogenic Global Warming 27

apples 52, 53, 175

aquifers 62, 114

Archangel 64

Archibald, David 150

Archimedes 47

Arctic 39, 64, 65, 69, 70, 73, 77, 84

Arctic Circle 64

argon 14, 15

argumentum ad populum 28

argumentum ad verecundiam 28

Aristotle 28

Armstrong, J. Scott 89

Asch 29

ash, volcanic 154

asteroids 153, 219

astronomers 119

Atlantic 50, 58

atmosphere 5, 7, 9, 19, 21-26, 54, 55, 67, 68, 71, 80, 84, 88, 90, 95, 96, 99-101, 105, 112-114, 119, 121, 127, 128, 135, 142, 148, 154, 155, 184, 188, 189, 192, 214, 223

atmospheric hotspot 2, 90

atolls, coral 51

atomic number 13

atoms   5, 13-16, 52, 53

atonement 134, 136, 139, 158

Australia   31, 39-41, 43, 60, 69, 76,
    110, 126, 127, 155, 156, 164, 187,
    192

Australian Coal Association 127

Australian Coal Industry   69

Australian Government   69

Avatar   153

bacteria   19, 25, 26

    anaerobic   25

    oxygen-hating   26

Baffin Bay 64

Baluchitherium   22

Ban Ki-moon   32

Bangladesh 50, 51

bats, insectivore   156

battery acid 52, 53

BBC43, 142

Beck, Ernst-Georg   112, 113

bed bugs   165

Beenstock   108-110

beer 52

Bering Strait   64

Big Coal   11, 126-128, 140, 196,
    197

biodiversity 147, 148, 158, 191

biofuels   39, 174, 175

biology   88, 148

biomass   182, 183, 185, 194

birds   21, 22, 132, 134, 145,
    153, 156, 157, 164, 166, 167, 170

blood   52, 156

bourgeoisie 219

Brachiosaurus   21

Brisbane   41, 103

bristlecones 80, 82, 83

Bubonic Plague   42, 165

Buddha   212

Burnett, H. Sterling 64

bushfires, Victorian 156

C3headlines   193, 194

CAGW   27, 221, 227

calcium   52, 55, 189

calcium hydroxide   52

California 129, 166, 167, 171, 182

Cambrian   20, 25, 45, 108

Cambrian explosion 20, 25

Cameron, James   153

Canada   56, 57, 63, 64, 66, 80,
    141, 152, 164, 179

cannibalism 216

Canute, King   123

cap and trade   11, 125

capitalism 139, 205, 209

capitalists 139, 140

carbohydrates   16

carbon

    buried   26

    credits   125

    cycle   9, 17

    donation 9

    footprint 8, 9, 24, 135, 138,
        140, 170, 179, 188, 190, 192,
        226

    market   140

    pollution 8, 17, 192

    starvation 9

carbon capture and storage   128

Carboniferous   20, 119

Carson, Rachel   164, 167

Carter, Robert M   112, 225

caustic soda 52, 53

CCS 128

Cenozoic   21, 22

censorship 177

Center for the Study of Carbon
    Dioxide and Global Change
    187, 188

CERN   116-118, 120

Chagas Disease   165

cheetahs   143

chemistry   13, 16, 52, 88, 111
    organic   13

Chicago Field Museum   184

child labor 174

China   173, 183

chlorine   14, 15

chlorophyll 17

Choi, Yong-Sang   103

Christmas   42, 166

Chukchi Sea   64

CITES   66

civilisation 3, 113, 117, 148, 149,
    153, 201, 207, 214, 217, 219, 228

clean coal   11, 191

climate   2, 3, 5, 10, 11, 27-37, 39,
    40, 42-45, 55-57, 60, 61, 67-69,
    73, 74, 77, 78, 80-84, 87-91, 95,
    96, 98, 99, 101-105, 108-110, 112,
    113, 115-120, 123, 125-129, 136,
    141, 143, 144, 147, 149-151, 154,
    155, 159, 163, 169, 170, 173, 174,
    180, 184, 186, 188, 190-192, 194,
    195, 206, 215, 221-223, 225, 226,
    228
    change   10, 27-29, 32-36, 43,
        60, 84, 88, 89, 104, 105, 117,
        125-129, 136, 141, 163, 169,
        173, 174, 186, 188, 190, 191,
        221, 225, 226
    change deniers   10
    crisis   2, 143
    models   2, 57, 60, 88, 90, 91,
        99, 101, 102, 105, 108-110,
        120, 150

Climate Audit   80, 82, 223

climate challenges   29

climate realists   37, 186

climate refugees   30

climateaudit.org   82, 196, 223

Climategate 2, 29, 130, 131, 226

Climatic Research Unit   76,
    129

Clinton, Bill   141

Club of Rome   137, 190

clusters, open   120, 121

Colorado, University of   49

common enemy   137

compounds 13, 14, 156, 185

Congressional Budget Office 126

consciousness   79, 133, 134,
    214

conspiracy 2, 130, 163

continental drift   20, 121

Convention on International Trade

in Endangered Species   66
cooperation   137, 138
Copenhagen   65, 130, 131
Copenhagen Consensus Center
    65
coral   30, 36, 41, 51, 54, 55,
    71, 222
cosmic rays 2, 115, 116, 118, 120,
    121, 123, 154
Cretacious 21
cropping   152
crops   9, 11, 23, 57, 62, 155,
    174, 181, 184, 194
CRU   76, 78
Crutem4   78
cultural revolution   137
cyanobacteria   19
D'Aleo, Joseph   74, 75
dark age   59, 205
Dark Ages 35, 42, 179, 205, 206
Darwin   5, 76, 77, 118
DDT   3, 64, 164-167, 169, 171-
    173, 201, 206, 214, 216, 221
decibels   53, 54
deforestation   114
Dengue Fever   165
descendants 189, 210
Devonian   20
dinosaurs   20-22, 24, 71, 119, 153
Diplodocus 21
disadvantaged   12
disasters   30, 41
diversity   118, 121-123, 182
dodos   157

drought   5, 23, 60, 61, 84, 185,
    217
droughts   1, 30, 31, 41, 60, 122,
    181
Dyson, Freeman   88, 89, 106, 184
eagle   101, 166
Earth   1, 2, 5, 7-11, 17-21, 24-
    26, 31, 33, 39, 44-46, 55-57, 60,
    65-68, 75, 78, 85, 91, 95, 96, 100,
    102, 103, 105, 106, 108, 109, 115,
    116, 118, 121, 123, 125, 131, 133,
    137, 138, 141, 144, 145, 148-151,
    153-155, 160, 161, 163, 164, 182,
    186, 188, 190-192, 195, 197, 207,
    210, 215, 218, 219, 224, 225
Earth Day   164
East Anglia, University of   75,
    130
Easterbrook, Don J 44
eco-environmentalists   190,
    217
eco-religion 192, 204
ecology   88
ecosystems 114, 155, 181, 186, 202,
    219
Ediacaran fauna   19
Edwards, J. Gordon 166, 167
Eemian   22, 44, 55, 65, 71, 78,
    144, 146
Einstein, Albert   136, 184, 196
El Niño/Southern Oscillation
    115
electric charge   13
electron   5, 13-16

elements   1, 2, 6, 9, 13-17, 19, 28, 141

Elephantiasis   165

elevator experiment 29

Elizabeth I 208

energy   3, 5, 7, 8, 17, 25, 37, 39, 40, 54, 58, 59, 69, 78, 91, 98, 100, 101, 103, 110, 113, 115-117, 119, 123, 127, 128, 149-152, 169, 171, 172, 174, 189-191, 194, 215-217

engineering 152, 203

English   63

Enlightenment, the 205

environment   8, 10, 11, 26, 28, 31, 38, 44, 46, 65, 113, 114, 134, 136, 137, 139, 141, 154, 158, 159, 176, 181, 202, 204, 210, 226

environmental movement   156, 157, 197, 198

environmentalists   80, 138, 140, 190, 197, 217

Eons   18-20

eras 20-22, 27, 42, 54, 75, 181

Eschenbach, Willis   107

ethanol   3, 175, 193, 194

Etna154

euphemisms   158

Europe   40, 42, 43, 63, 64, 84, 116, 126, 159, 164, 188

European Commission   129

Everest   32

evolution   21, 26, 44, 63, 118, 123, 143, 201

extinction   20, 21, 26, 143, 154, 156, 160, 166, 171, 180, 189, 191

extinctions 11, 30, 157-159

extreme heat   78

extremophiles   19

Falagas, Matthew   43

falciparum 64

famines   122

farms   23, 39, 88, 128, 156, 158, 170, 171, 187

fertiliser   11, 194, 214

fish 21, 31, 32, 156, 157, 202

floods   1, 32, 41, 51, 60, 122

Florida   32

fluid dynamics   88

flying foxes 156, 170, 171

food7, 9, 10, 17, 23-26, 32, 33, 58, 71, 102, 110, 111, 122, 123, 127, 128, 136, 141, 147-149, 152, 154, 160, 163, 172, 174, 176, 181, 184-194, 199, 214, 217, 223, 228

    for fuel   9, 194

    prices   9, 32, 174, 181, 194

forests   8, 57, 88, 114, 155, 180, 185, 186

fossil fuel   55, 101, 127, 174, 176, 194, 228

Fourth Assessment Report   79, 81, 89

fraud   9

free market 139, 176

freeway   134

Friedman, Richard   206, 207

friendship 138, 142

funding   69, 196, 197

fungi   20, 31, 32

Gaia39

galaxy   119, 120, 154

Gandhi   212

Garden of Eden   133, 219

gases

   noble   14

GDP   174, 206

Genesis   133

GISS   49, 75

Giza44, 45

global climate disruption   29

global warming   1, 3, 8-11, 27-
   37, 39-42, 44, 46, 47, 49, 51, 56,
   59-61, 63-67, 69, 75, 78, 84, 91,
   95-99, 102-104, 106-111, 116,
   117, 125, 126, 128-131, 136-142,
   144, 146, 148, 155, 156, 158-160,
   164, 168-170, 172, 173, 177, 179,
   180, 184, 186, 190, 193, 195, 198,
   201, 216, 221, 222, 224, 225

glucose   8, 17

God 192, 204-208, 211, 225

Goddard Institute for Space Studies
   75

Goddard, Steven   54, 75

Goklany, Indur M.   173, 174

good and evil   213

Goodness, Principle of   1, 215

Gore   36, 80, 146, 179, 197,
   226

government

   global   131

governments   8, 66, 69, 125,

129, 209, 210, 226

grass7, 8, 22, 24, 147, 148, 181-183

Great Pyramid   44, 45

greater good   177, 211, 216

Green, Kesten C.   89

greenhouse 2, 23, 45, 91, 95, 99,
   100, 102, 105, 108, 109, 174, 181

Greenhouse Gasses 2, 99

greenhouses   2, 9, 23, 45, 91,
   95, 99, 100, 102, 105, 108, 109,
   174, 181

Greenland 42, 47, 48, 57, 84, 108,
   179

habitats   1, 23, 55, 56, 65, 171,
   202, 210

Hadean   18, 19

Hadley Centre   74

Happer, William   136

happiness   198, 203, 207, 209, 213-
   215, 228

hardships   1, 23

Hayek, Friedrich   139

helium   13, 14, 16, 68

Heuer   117

hindcasts   106, 107

history

   geological   11, 145, 149

Hitler   164, 168

hockey stick   1, 5, 35, 79-83,
   223, 226, 227

Holocene   22, 44, 45, 55, 65, 70,
   144, 151

homo sapiens   133, 191

hotspot   2, 5, 6, 90, 96-99, 102,

105

House of Lords   63

HSBC Bank   129

humanity   7, 10, 11, 39, 44, 46,
    127, 133, 134, 136-138, 142, 164,
    166, 167, 174, 179, 191, 192, 202,
    207, 210, 214, 215, 218, 228

hunger   174, 175, 194, 201

hurricanes   1, 58

hydrochloric acid   52, 53

hydrogen   5, 13-15, 17, 52-54, 68

ice 2, 6, 7, 10, 21, 22, 24, 32, 35, 42,
    44-49, 55-57, 59, 63, 65, 67, 69,
    70, 73, 74, 77, 79, 84, 108, 115,
    119, 121, 142-152, 155, 160, 179,
    180, 186, 192, 201, 223
    Antarctic   21, 48, 121

ice age   2, 10, 21, 22, 24, 32, 35,
    44, 45, 48, 55, 56, 59, 63, 65, 67,
    70, 74, 79, 84, 115, 119, 143-152,
    160, 179, 180, 186, 192, 201, 223

ideological beliefs   132

ideologies   139, 172, 173, 177, 192,
    202, 204, 205, 209, 216, 218

Independence, Declaration of
    209, 210

India   205, 217

industrial revolution   10, 23

insects   20, 156, 157

insolation   115, 149-152

insurance   163, 169, 177

interglacial 22, 36, 44, 55, 65, 71,
    78, 144, 146, 151, 201

International Conference on Rising

Atmospheric Carbon Dioxide
    and Plant Productivity   187

internet   14, 30, 133

ion, hydrogen   53, 54

IPCC   28, 29, 56, 79, 81, 82,
    87-90, 95, 99, 106, 125, 128, 129,
    141, 226

Japanese Encephalitis   165

Jefferson, Thomas   210

Jesus   80, 211

Jung 198

Junkscience.com   172

Jurassic   21, 24, 71

Katrina   36

Kenya   40

Kilimanjaro   36, 84, 155

knowledge
    scientific   8

Krakatoa   154

land fill   135

language   54, 78, 132, 136, 219,
    224

late heavy bombardment   19

latitudes   73, 95, 186

laws
    physical   8

lemon juice 52

leprosy   165

liberty   209, 210

Lice 165

lie, big   168

life 1-3, 5, 7-12, 16-22, 24-26, 33, 41-
    47, 67, 78, 88, 92, 102, 108, 110,
    118, 119, 121-123, 125, 127, 130,

132, 133, 136, 139, 141, 142, 144-147, 150, 153-156, 160, 161, 163, 165-169, 172, 174, 179, 180, 186-192, 196-199, 201, 203-210, 212, 214, 216-219, 223

building block   7

cycle   7, 9, 17

life expectancy   174

life on earth   1, 18

light bulbs 136, 150

lime 52

Lindzen & Choi   103

Lindzen, Richard S 103

lion 133

logarithmic scale   53, 54

Lomborg, Bjorn   65

London   43, 73, 144

Lovelock, James   39, 40

Lymes Disease   165

Magnesia, Milk of 52

magnetic activity   115-118

magpie   1, 132

malaria   1, 3, 63, 64, 164, 165, 167, 172

males   219

malnutrition   174

Malthus, Thomas   190

mammals 20-22, 157, 202

Manchester 144

Mann   80, 81, 83

Mao 173

Mars   18, 67, 68

marsupials 21

Mary   208

mass extinction   20, 21, 26, 154, 180

Maunder minimum 115

McDonald's   153

McIntyre, Steve   80-83, 223

media   24, 29, 30, 33, 34, 44, 50, 59, 61, 64, 66, 69, 70, 84, 125, 126, 151, 152, 164, 168, 175, 186

medieval climate optimum   42, 78

megafauna 22

meme   64, 159, 228

mercury   136, 226

Mesozoic 20

meteorites 19

migrants 32

Milankovitch cycles 144

milk 52, 164

models   2, 9, 57, 60, 67, 87-91, 94-96, 98, 99, 101-103, 105, 106, 108-110, 120, 150, 157, 184, 186

molecules 8, 13, 15-17, 53, 67, 179

Monckton, Lord Christopher   28, 29, 36, 103, 131, 167

monotremes   21

Montford, Andrew 80-83, 226

Moon   19, 32, 60, 120, 150

Muhammad   211

Myers-Briggs test   198

NASA   5, 48, 49, 74, 75, 104, 105, 129, 144

NASA water vapour project 105

National Academy of Sciences   167, 226

National Climatic Data Center

Carbon is Life

10, 75

National Climatic Data Center
(NOAA)   10, 75

nature
  back to   134, 152
  documentary   138

Nature (journal)   184

NCDC   56, 58, 59, 61, 75

Neandertals   144, 146

Nemani, Ramakrishna R   185

Nepal   32

Newton   196

NIPCC   89, 96, 106

nitrogen   2, 14, 16, 156, 187, 202

nitrous oxides   114

NOAA   10, 56, 58, 59, 61, 74-76,
  129

Noble Cause Corruption   175

North Pole 95, 149

NOx   194

Numberwatch   30, 32

Nunavut Province 64

NVAP   105

oceans   1, 30, 47, 48, 50, 52, 54,
  55, 68, 88, 90, 99, 102, 104, 121,
  122, 150, 156, 193, 202

Open University   154

Ordovician 20, 54

organic molecule   8, 17

Oxfam   175

oxygen   8, 14, 16, 17, 19, 25, 26,
  53, 114, 156

ozone   114

Pacific Decadal Oscillation   115

Pacific islands   36

Paleoclimatology   112

Pandora   153

Pangaea   20, 21

passenger pigeons   157

peacelegacy.org   47, 170, 224

peer review 29, 164, 194-196

Pendon Museum   217

perfect storm   126, 140

periodic table   6, 14, 16

permafrost 42

Permian   20, 21, 180

personality 198

pH6, 52-54, 188

Phanerozoic   20, 22

Phillip, Prince   136

phlogiston 111

photosynthesis   8, 17, 19, 147,
  148, 160, 161, 187, 189

physics   2, 8, 13, 87, 89, 91, 95,
  98, 99, 103, 105, 106, 109, 131,
  150, 222

placentals 21

planet   3, 9-12, 18-22, 24-27, 30,
  31, 37, 41, 42, 44-47, 55, 57, 60,
  66-69, 71, 73, 74, 78, 80, 81, 83-
  85, 87-89, 95, 100-102, 104, 107,
  110, 113, 114, 116, 122, 123, 127,
  132-136, 140, 144, 145, 147, 148,
  150-153, 155, 158-161, 170, 177,
  179, 180, 185, 188, 190, 191, 193,
  194, 201, 203, 210, 214, 216, 218,
  219, 225, 226, 228

plant food 26, 33, 71, 110, 123,

127, 128, 160, 163, 172, 181, 191, 192, 228

plant growth 1, 3, 9, 23, 111, 160, 181, 184-187, 190, 192

plants 1, 3, 7-9, 11, 17, 19-26, 56, 58, 111, 123, 127, 128, 141, 147-149, 153, 155, 160, 180, 181, 183-185, 187, 189, 222

Pleiades 119

Pleistocene 22, 143, 201

Plimer, Ian 60, 78, 122, 224, 225

polar bears 1, 26, 30, 36, 41, 64-66, 70, 71, 123, 145, 179, 199
 photos 64

politicians 9, 24, 29, 33, 40, 44, 66, 99, 128-130, 140, 142, 165, 192, 196, 197

pollution 7-9, 17, 56, 129, 135, 137, 170, 185, 188, 189, 192, 202

population bottleneck 143, 201

population policy 24

population pressure 136

poverty 40, 41, 63, 78, 79, 173-176, 181, 203, 207, 208, 216, 218

power
 solar 127

Precambrian 19

precautionary principle 3, 39, 111, 163, 169, 177, 187

precipitation 5, 62

prosperity 7, 131, 143

proteins 16

proton 13, 14

pterosaurs 21

pyramid 44, 45

quantum mechanics 8

Queensland 190

R404A (refrigerant) 134

rainforest 10, 46, 155

red giant 68

Reiter, Paul 63

religions 173, 205, 206

Remote Sensing Systems 75

renewability 170, 190, 191

renewable energy 3, 169, 190, 191, 217

reproduction rate 159

reptiles 20, 157

resources 4, 11, 27, 41, 45, 66, 70, 127, 152, 153, 176, 190-193, 202, 222

Reykjavik 150

Richter scale 54

River Blindness 165

robin 166, 217

Rocky Mountain spotted fever 165

Roman 35, 42, 63, 71, 78, 179, 189, 206

Roman Warm Period 35, 42

Rome 137, 190, 205

RSS 75

Russia 63, 173

Rutledge, D 164-166

Sahara 10, 84

San Francisco 32

Sarker, Maminul Haque 51

scare   1, 3, 9, 69, 70, 103, 110, 125, 126, 129, 137, 142, 151, 157, 169, 172, 186, 189, 191-193

Schrödinger, Erwin 8

science   1, 8, 25, 28, 31, 34, 37, 40, 51, 52, 60, 69, 73, 78, 81, 83, 87-89, 94, 98, 99, 104-106, 111-113, 120, 125, 129, 131, 133, 136, 139, 141, 164, 165, 185, 186, 195-197, 203, 206, 208, 215, 224, 225

computer   1, 8

science is settled   125

scientific method   37, 106, 210

sea levels   1, 5, 30, 44, 47, 48, 50, 85, 90

Self, Stephen   154

sentient beings   158, 218

separation   133, 134, 137, 219

Seven Sisters   120

sharks   20, 21

Shikwati, James   40, 41

Siberia   10, 57, 63, 64, 179

Silent Spring   164, 166, 167

silicon   14, 16, 17

Silurian   20

Simberg, Rand   83

Socrates   196, 212

Sodium   14, 15, 52

Sodium Hydroxide 52

solar radiation   186

solar system   19, 120

Sophistical Refutations   28

Spain   144, 197

species   7, 9, 11, 12, 20, 24, 26, 32, 42, 44, 46, 48, 55, 56, 58, 66, 80, 114, 123, 132-134, 136, 143, 144, 146, 147, 154-158, 171, 172, 179, 180, 182, 183, 189, 191, 201, 202, 205, 206, 219

spontaneous generation   111

St Helens   154

St. Louis Encephalitis   165

Stalin   173

star-forming region 119

stars 118-121, 154, 212

starvation   9, 23, 74, 154, 159, 176, 184, 189

Stephens, Bret   129

Stewart, Christine   141

Stocking, Barbara   175

Stonehenge 44, 45

storms   1, 30, 58-60, 122

sugars   16

sulphur dioxide   114, 192

sulphuric acid   52

summer   43, 46, 57, 65, 78, 84, 108, 145, 149-151

Sun 7, 17, 67, 68, 100, 115-121, 123, 149-152, 154, 189

magnetism 116

sunspots   115

supercontinent   20, 21

Supernovae 2, 66, 100, 118, 119, 121, 154

supervolcanoes   66, 74, 154

surfacestations.org 77, 223

sustainability   216

Svensmark, Henrik 2, 116-118, 120-

122, 154

swamps   20, 88, 114

Syracuse   47

Taliban   31

technology 40, 41, 126, 127, 136, 152-155, 188, 203, 205

telescopes   115

temperature   1, 2, 5, 6, 10, 11, 34, 36, 44-47, 55-59, 63, 64, 66-68, 73-78, 80, 83, 84, 88, 93-100, 103, 105, 107-110, 114-117, 121, 122, 125, 131, 145-147, 149, 155, 160, 163, 180, 186, 223, 225-227

thermal   27, 101

Thermohaline circulation   36

thermometers   76, 77, 83, 227

thermostat 67, 68, 103

Third Assessment Report   79

Thurow, Lester   205

thylacines 157

Tiger Woods   137

timber   155

tipping points   1, 27, 66-68

tomatoes   52

toolmaker 132

transportation   152

Triassic   21

Triceratops 21

trilobites   20

typhus   165, 167

Tyrannosaurus Rex 21

tyranny   131

UAH   75

UNEP   28

United Kingdom   128, 154, 183, 187

United Nations   28, 137

United Nations Environment Program   28

unity   137, 138

universe   8, 16, 67, 88, 204

University of Alabama   75

uranium   13

USA   56, 59, 128, 183

useful idiots   197

Venus   31, 67

Vesuvius   154

Vietnam   164

Viking   42

Vinegar   52

Wahl and Amman   81

Warmlist   30, 32

warmth   1, 7, 31, 42, 43, 46, 64, 65, 73, 102, 121, 125, 145, 148, 150-152, 160, 169, 180, 189, 192

Washington   26, 63, 81

water   7-9, 17, 19, 23, 37, 47-49, 51-54, 60, 62, 67, 68, 84, 99-105, 114, 117, 121, 137, 146, 149, 152, 156, 181, 187-189, 194, 202

acidic   19

Watts, Anthony   74, 75, 150, 151, 222, 223

wattsupwiththat   28, 29, 40, 42-45, 54, 60, 64, 76, 78, 99, 105, 107, 112, 116, 129, 142, 150, 174, 184, 195, 222, 223 weather   2, 30, 32-34, 41, 42, 58, 59, 65, 83, 84, 90, 122, 126

West Nile virus   165

Westerners 188, 193, 194, 214, 219

Westminster   63

whites   219

whooping cranes   172

Wiegand, Jim   170

Wikipedia 157

wilderness 9, 26, 176

wildlife   1, 3, 8, 9, 23, 24, 58, 63, 64, 66, 71, 74, 111, 114, 123, 127, 128, 133, 136, 155, 156, 158, 160, 163-165, 170, 181, 190, 193, 214

wind energy   113, 171, 172

wind farms 39, 128, 156, 158, 170, 171, 187

wind turbines   3, 170, 171

wine   52

Wirth, Timothy   141

wisdom, spiritual   172

WMO   28, 74

world government 131

World Health Organization 172

World Meteorological Organization   28

WWF   64, 65

Yellow Fever   165

Yellowstone National Park   154

Zichichi, Antonino 106

www.ingramcontent.com/pod-product-compliance
Lightning Source LLC
Chambersburg PA
CBHW051450170526
45166CB00001B/189